Electronics Constructors Guide

To my father, Dennis Cuthbertson, and to my friend, David Beale. Unfortunately, neither of you are here now to see this book, but your dedication and craftsmanship have influenced me tremendously.

Thank you.

When we build, let us think that we build for ever.

John Ruskin, *The Seven Lamps of Architecture*

Electronics Constructors Guide

Paul Cuthbertson

Newnes
An imprint of Butterworth-Heinemann Ltd
Linacre House, Jordan Hill, Oxford OX2 8DP

A member of the Reed Elsevier plc group

OXFORD LONDON BOSTON
MUNICH NEW DELHI SINGAPORE SYDNEY
TOKYO TORONTO WELLINGTON

First published 1995

British Library Cataloguing in Publication Data

A catalogue record for this book is
available from the British Library

ISBN 0 7506 2211 3

Library of Congress Cataloguing in Publication Data

A catalogue record for this book is
available from the Library of Congress

Printed in Great Britain by Clays, St Ives plc

Contents

Preface

Some years ago it became apparent to me that, despite a great deal of effort on their parts, some of the people I was helping experienced great difficulty in understanding what was needed to get electronics projects into a working, tidy state. Projects seemed to vary between the magnificent and the totally unworkable, with a spectrum of all variations and combinations of tidiness, completeness and robustness in between.

An actual working project is not always a definite requirement for, say, degree students (there are other, equally important, issues at stake). Very often, however, a project would be put about as working, which in the event proved not to work at all, excepting perhaps under certain tight constraints, rendering it pretty useless except as an idea for further development.

It distressed me to see so much effort result in something which, with a little effort applied in certain ways, would render the device workable. At the very least, parts would not fall off and wiring disconnect itself under the strain of being moved from one bench to another.

It also struck me that certain kinds of approaches succeeded in producing proper working instruments and other approaches did not. One of the big differences between a working project and a non-working project was that of organization. Not the regimentation of the production line, but imaginative planning and the imaginative use of time, materials and resources.

Having allowed these thoughts to mature over a further half a decade or so, and using the techniques as I do myself from day to day, I thought it prudent to commit them to paper before they were either lost forever or before events otherwise overtook them. So here they are.

Paul Cuthbertson, Banchory 1994

Acknowledgements

Thanks to the staff at Butterworth-Heinemann for their support and advice in producing this very first book.

Thanks also to Texas Instruments Ltd. and National Semiconductor Corporation for their kind permission to reproduce parts of their data books.

Thank you, too, to the many people who, over the last eighteen years or so, have provided me with the scrapes, experiences and opportunities to learn, both about electronics and about human nature. I have mentioned no names; both the innocent and the guilty have been protected.

Errors and omissions are all my own.

There are several people who offered to read this manuscript before publication and give me their comments, who unfortunately did not get the chance due to deadlines. You know who you are; perhaps I can offer my thanks in advance as you read the finished product. I hope you like what you see; your comments would still be very valuable!

Particular thanks to Mollie, a good friend possessed of an almost inhuman quantity of patience.

And finally, all Registered Trade Marks used within this book are hereby acknowledged.

1: Introduction

Scope

This book covers most aspects of the design and construction of small to medium sized electronics projects including d.c. to low radio frequencies and involving modest currents and voltages. By modest, I mean mains voltage or less and perhaps less than ten amps of current.

These limits have not prevented me from including hints for those thinking of higher voltage or current limits where these are simple extensions of the examples given.

I have tried to avoid treating design as a 'subject' for academic study, but rather as a series of appropriate techniques, with examples where needed. I hope that you will quickly develop a feel for proficient design. There are hints and tips of all kinds to assist in getting a good design off the ground with the minimum of fuss and red tape.

My intention is that good design decisions shall be as easy for you to make as possible.

Who is this book for?

The book is initially written for the enthusiastic amateur or student, although it may equally well be used by the professional engineer whose experience of electronics design has hitherto been limited and who needs a good introduction.

My own particular experience with students suggests that they don't always use good partitioning or productive test and measurement techniques.

School students may very well benefit from studying the chapters on design principles, particularly the examples.

If you have a bright idea in the offing and you are keen to make it work for you in the long term and on a large scale, then this book is particularly for you.

What will you learn?

Segmentation of design, both from the point of view of timescales and planning and partitioning of the actual electronics itself, is a recurrent theme. 'Deferred design' is introduced too.

I have included a good few hints on designing for usability, testability, maintainability, adjustability and ruggedness. Too often, a project is just a board with a heap of wires sticking out of it. With very little more attention to detail at the outset, such a beast could have emerged as a thing of beauty, perhaps even lucrative, commercial beauty.

Those things ignored by most pundits as being 'not electronics' have not been dismissed and there are extensive sections on such things as the mechanics of electronics, choosing enclosures and wiring, cooling, panel layout and labelling. These kinds of things are too often left to guess-work or 'we'll see how it fits into the box later', i.e. after we've finished all the interesting soldering bits.

Throughout, there is an emphasis on keeping documentation up to scratch, but not to the exclusion of working on the actual project itself. Documentation has its uses, even for the domestic constructor, and I have tried to make it easy for the various classes of reader to establish their own sensible needs for documentation.

Documentation is perhaps one area where a simple formula is appropriate as a starting point for generating your own documents. Complete, useful documentation in a sense depends upon the application of various check-lists, so I have used a 'cookbook' approach in this case.

Subjects such as these may not interest those designers who are just knocking up a little gadget for their own use. However, I think it is important to factor into *any* design, where this is easy to do, those things which might make the difference between a robust, generally interesting and useful device and yet another 'thingummy with wires' that our Paul made, and which now gathers dust at the back of a drawer (yes, I've made a few of those in my time too).

In Chapter 6, which gives a broad overview of the available electronics devices and suggests the manner of their use, I have very much treated components as 'black boxes'. That is to say, I have not dug into their innermost nature to any great extent but have instead concentrated on their behaviour. From a systems viewpoint this is perfectly adequate; we are in fact building a hierarchy of complexity, of which the individual component itself is the lowest useful level.

This book cannot demonstrate all possible uses of any given component. It is intended to be neither a source book of circuits nor a cookbook, although these definitely have their uses and are in fact well represented on my own shelves.

Information on components is easy to come by. I have resisted the temptation to duplicate such information at length. Rather, I have given some hints on its intelligent use.

Those circuit and systems examples which I have used are mostly taken from real situations which I have encountered. I have often examined them in minute detail, not in any way to enhance an understanding of the circuits themselves (which are very simple in most instances) but to highlight the approaches and thought processes which can identify problems or potential problems and which quickly yield acceptable solutions or preventatives.

I hope that those chapters dedicated to design will leave you with the idea that all is fair, provided that any assumptions are thoroughly tested so that the finished article obeys all our previously mentioned criteria of constructibility, usability, completeness and ruggedness.

Those parts of the book devoted to construction go on to give details of the advantages and pitfalls of the major construction techniques readily available today. Correct use of tools is covered, but more than this, you are encouraged to use inexpensive tools imaginatively and appropriately in order to avoid unnecessary expense. The uses of prototyping are mentioned and segmentation is reviewed in the light of the experience of prototyping.

Lastly, there is a chapter devoted to production. There is, I hope, an obvious, continuous thread through the previous chapters leading up to this point. If you have imaginatively used the hints I've given in the rest of the book, you'll be well-positioned to succeed in getting your favourite project out of the door and into the market-place. I also hope that a great many of you will have a go at it.

There are some things which are definitely not covered in this book. Involved mathematical treatments are out. Often, mathematics is used to prove a point or to attempt to describe some phenomenon after the fact. To get something working, we need do neither of these things.

Do not get the idea, however, that I am anti-mathematics; far from it. Math is a beautiful way of expressing ideas and succinctly representing what would otherwise be impossible to describe. For us, however, mathematics will be used purely as a tool for exploration and prediction of the behaviour of simple circuitry. For all the examples given in this book, simple algebra will suffice to get the right component values. Then we need to use our wits to select a real component whose value or behaviour is close enough for our needs!

That most fundamental of relationships, $V = IR$, is given a thorough airing. I believe it to be the single most useful piece of theory, applicable to almost any situation in electronics. It is truly remarkable how many people think that Ohm's law is not useful for their particular problem, or that it somehow varies with the weather, the phases of the moon or which way up the circuit diagram is. $P = IV$ is given similar treatment. Even if this book sets only these simple facts straight, I think it will have been a great service to humanity!

Our justification is the end product. The questions we are seeking to answer are: does it work; can I get the bits easily; will its fellows work if we build a dozen with the same component tolerances; and will it carry on working for a good while?

Software has not been mentioned except as it impacts on the areas of electronics design, computer-based digital electronics being the prime example of course, where some trade-off is to be expected between the complexity of the hardware and the software and in which the design boundary can be moved one way or the other, whilst still leaving a similarly functional (so far as the user is concerned) system. Otherwise, discussions of software are confined to those design aids which are likely to be useful.

Occasionally I have lapsed in to psychology. Design is a human activity carried on by human beings, and the end result of the design process is the production of some article intended to be used by human beings too.

I make no apology, therefore, for attempting to alert you to the possible uses of your own powers of imagination and observation and the reactions of people when faced with the need to use high technology. Good design takes note of the psychology of perception, cognition and learning. Fortunately, the rules are fairly simple.

Sometimes, despite our best efforts, things will go wrong. In that kind of situation, despair is not an option. Rather, corrective action is the course to take. If all the planning is in order and documentation is available and legible then it is always an option to backtrack, to find the point of failure and make good. Expect failure by all means, but expect to succeed in the end.

Above all, I want to bring home the need for good judgement, choices and decisions. There are an infinity of possible ways of doing the same thing, even in the simplest kinds of design. Even choosing to abandon a design is a choice, not a default. The ability to choose wisely from a vast array of possibilities, to achieve sensible compromises and gain optimum benefit in terms of effort, cost and time is a real skill. But like all skills, it can be learned.

Design and construction phases: overview

It seems trivial, in a way, to mention that design comes before construction, and yet it is remarkable how often the two go on side by side. This may be due in some situations to the terrible temptation to begin soldering as soon as possible, in the vain hope that the sooner we start the 'real work', the sooner we will be finished.

On the other hand, it may be that some powerful plan is in action, which allows us to get on with various parts of the project in the full knowledge that when the time comes to fit the individual pieces together, they will indeed fit together properly, because there is some previously agreed set of standards with which the individual units comply.

In a well planned project, some aspects of design and construction might easily go on side by side. In all but the simplest projects, several activities can be under way at the same time.

Inevitably, however, there is some element of design at the outset. At the very least, someone will have decided that such-

and-such is going to happen and they will have a more or less complete mental picture of how it will look in the end. The final outcome will be affected by how complete that picture is, how well it is communicated to the rest of the team or to a superior and the quality of influence (or interference!) which the picture suffers, both at the hands of colleagues and the constraints of reality.

I like to think that, ideally, design and construction consist of firstly a proliferation of ideas and a flurry of planning and partitioning ('segmentation'), which settles into a resolution phase (during which the main work of design and testing is done) and then finally a kind of tidying up and overall testing phase.

Within this major project evolution, minor design questions, spawned from the main design thrust by good partitioning at the outset, will each evolve along similar lines to the parent, merging back into a coherent whole at the end. Each minor phase may spawn lesser issues and so on, until we get to the level of component choice. The business of recursive partitioning and its related issue, deferred design, are discussed in practical detail later.

This progress model seems to suit electronics particularly well. In particular, tidying up and testing neatly round off each minor design phase of a good project plan; testability is a good feature of most electronics.

This testability is a peculiarity of the game and a key, I reckon, to the success of electronics as a technology. Other engineering disciplines may not be so lucky. Railway bridges, for instance, are not the most eminently testable of items, at least when partially built. Engineers must resort to testing materials and components before inclusion in the finished article (a good ploy in any engineering discipline really) and to modelling the finished product either on a computer or physically on a small scale.

All of this takes time and effort and is expensive compared to the breadboard trials and subsystem testing which are available to us in electronics. In electronics, we do not have to wait for the whole system to be built before we can start rigorously testing individual parts. Breaking up an electronic system into component parts is fairly simple.

Amongst the closest allies of the electronics engineer, software engineers are plagued by a lack of testability, and are forever coding their own 'instruments' to check on the performance of their other, 'real' program code. They are also up against the wall due to the lack of isolation of a piece of software, which will almost always share the computer system with some other potentially interfering programs, over which they have little or no control. Shame! Electronics is, more often, easy to isolate and to feed with whatever inputs we think fit.

It is a pity, therefore, when the testability of electronics is so good, to see someone build a fairly complex electronic system for which a simple testing procedure has not been worked out, and which ultimately fails to come up to expectations because the design was not partitioned into easily isolated, testable modules, and then that testing actually done. It is often fatal to a project to build the complete thing without checking out the individual fragments as they become available for testing.

Occasionally we may come across systems which are less testable than usual. Computer systems are a case in point; anything attached to the common bus which fails is difficult to isolate from its fellows. Additionally, due to this close-knit interaction, the function of the computer as a whole is impaired by substitution of test components or removal of a possibly offending chip. Individual peripheral cards are not so awkward to test as the core computer, as they can be isolated on a test rig and exercised on their own.

The other major testability issue in electronics nowadays is that of the more complex integrated circuits (i.c.s), whose internal workings may be hidden and some of whose functioning is not available at the pins. These issues need not concern us too much here, however.

There are essentially two ways to construct your project. You can design and test everything and then go on to the construction phase afterwards. Or you can build fragments of the project as each is tested and verified as working. Exactly how all these activities fit together is discussed at length in the next section.

2: Design Activities

Good organization is essential to your design effort. Without it, you will find it difficult to work on your design effectively or communicate that design to your colleagues or other interested parties.

If this section seems a little academic, I would ask you to bear with me through my presentation of the models, as I hope you will soon begin to see the practical consequences of the ideas they contain.

Alternatively, come back to this section later. You can streamline the ideas contained here to make them fit your own uses better, but ignore them at your own peril.

Who's going to use it?

This is the primary question asked, or not asked, at the start of any project. The reason for not asking is that, usually, we think that we know the answer already.

I cannot over-stress the importance of putting yourself in the position of a potential user at the start of any project planning. Nor is it sufficient, unless you really intend the finished article just for yourself, just to imagine yourself using it.

Perhaps the best way to sound out potential users is to ask. This has to be approached carefully; you're not selling anyone anything. Concentrate on the usability issues rather than the internal technical features. Invite the user to imagine the instrument in use, but in a fairly impersonal way. It can be handy to have them imagining their colleagues using it or other members of their profession using it.

In the special case of having an actual customer, you have, in a sense, got a captive audience. Any such customer, having their wits about them, will already have well defined requirements and will let you know what it is they want right from the start, unless part of your brief is to do the market research too!

Sometimes this requirement may be short-circuited if you come into an existing project after the preliminary work has been done.

Hopefully, in this case the usability issues will have been resolved by the main contractor and you can concentrate on the engineering and actually getting the thing working. If both you and the existing project staff are communicative then you should be able to jump on board with the minimum of fuss.

We can divide users into two basic categories.

On the one hand, there are those who know what is going on inside the box and who might be fairly lenient on support and usability. Even so, don't abuse their tolerance. You, yourself, are the prime example of this class of user; after all, it's your design! Don't make assumptions about the knowledge of colleagues, unless they're engineers too; treat them like the intelligent people they are but recognise that they will need some support in the use of your system and will probably enjoy being kept up to date.

On the other hand, there are those users who do not have a clue what's going on under the hood and couldn't care less. That doesn't necessarily mean that they have a cavalier attitude towards your engineering skills and prowess, it probably just means that they have other things to do and they're relying on you to go out and make it right. These people may need a little guidance in specifying the user interface. It's not unknown for someone to specify, purely through inexperience, awkward or unworkable controls.

There are hundreds of firms who have gone out of business because they concentrated on what they wanted to build rather than what the user needed from them. Their numbers are as the grains of sand, as the stars in the sky. The basic rule is, if you're doing it for someone else, find out what they want. If they don't want to tell you what they want or they can't be bothered telling you what they want, then pull out. It's not worth it in the end.

Even if you have free rein, keep users or customers informed of what you're about, at an early stage, so that differences in perception and outlook can be resolved before the project gets to a point where major reworking is needed.

Remember that any underlying technology is subordinate, within the bounds of reasonable implementation, to the needs of the people who will use your work. No news is good news; if the system is invisible and inaudible, except when it needs to warn the user of some state of affairs or requires some input, then it probably means it's doing its job.

Take a minute to examine your motives with regard to a particular project. You may be undertaking the project: to make money; out of curiosity; for the challenge; because of a need to create; as a learning exercise; as a favour to a valued friend; because you're required to as part of an employment; because there is a need for the end product; or as a pastime. Perhaps there is some combination of these motives and maybe others.

All of these are fine and worthy reasons for going ahead; projects do not always need to be justified in terms of cash or usefulness. But don't kid yourself that you're doing it for the money when you're engaged in blue sky research, for example. Such a confusion of motives can corrupt the design process terribly.

Check out the motives of the people you're doing it for too. The point of the last two paragraphs is this: make sure everyone involved is on the same wavelength before you start. If motives change, be aware of it.

How long should it take?

There is an old saw, along similar lines to Parkinson's Law, called the 80-20 rule. It says that 80% of a job takes 20% of the time and that the other 20% of the job takes 80% of the time. Note, it is not always the *first* 80% of a job which takes the *first* 20% of the time, although it often works out that way since we might tend to rush ahead and do the 'good bits' first.

The 80-20 rule seems to hold quite well and is a good point to bear in mind when estimating for a job. It's too easy to gloss over difficulties and minor hiccups and to produce an estimate based on the first 80% of the work.

Most of the difficult or time-consuming 20% of the job is to do with making decisions which have to do with compromises for which there are no stock solutions. This is an iterative, or looping, part of a design effort. It's a thankless task going around in circles, working around constraints, and it may seem sometimes as though there isn't a good solution of any kind.

But be persistent. Psychologically, it's a vulnerable time; many of the poor decisions in design, I believe, have been made due to trying to cut corners during this 'difficult fifth'.

One excellent way of cutting down the time for this last 20% is to make sure your documentation is up to scratch. This is a time-saver in the long run and it is addressed in detail in further sections. Keep things tidy and in order too; frustration levels are minimized by having things to hand when needed.

Bear in mind that it takes 100% of the time to do 100% of the job. This isn't a re-statement of Parkinson's Law, which states that it takes 100% of the *available* time to do a job. It is merely a re-statement of the obvious which is often overlooked.

Within reason, minimize the unknowns. An estimate is a partial guess, but it can be a better guess by factoring in as many of the well-defined items as possible. Be realistic. Trying to get work by quoting unrealistically can push you under.

One way of postponing the awful moment of committing to a price and delivery timescale is to use 'incremental delivery'. This is more appropriate to the software scene but it can be applied, with care, to electronics or to design effort generally. Incremental delivery presupposes that your customer is willing to pay for fragments of the design without really knowing what the ultimate cost or delivery date will be or even, sometimes, if the concept is ultimately feasible; a 'cost-plus' situation beloved of suppliers to the military.

To use incremental delivery, you must agree with your customer, at the outset, what the end result of each phase of the project is, in general, to be. You must then contact the customer at these pre-arranged intervals to report in, deliver or demonstrate any existing fragments of the system and estimate the likely cost and timing of the next phase before going ahead with it.

If, during the early phases of the project, it becomes apparent that there will be difficulties or unjustifiable costs later, then the project can be abandoned with little harm done. The project will not have proliferated, taking up valuable resources, beyond a useful point. The 80-20 rule still applies in miniature to each incremental phase of the project.

Where learning curves of any kind apply, where innovative ideas are being tried out, be prepared to spend time. Innovation takes longer than the application of stock solutions, which is why there are so many of these. Be prepared to abandon innovation in those situations where it is shown to be more effective or more efficient to use ready-made solutions. Which brings us onto the next subject.

Effectiveness and efficiency

Let's make a clear distinction between effectiveness and efficiency. These two are quite different and there is a trade-off between them. For something to be completely effective it may need to be very wasteful. Similarly, for something to be as efficient as possible it may very well need to be partially, if not substantially, ineffective.

A streamlined and optimally-efficient delivery service, for example, will not reach every potential customer at all times, but will achieve the most deliveries for the least cost. It will do this by not delivering to out of the way places at all, or perhaps by bundling and delivering less often.

A totally effective delivery service will always deliver on time to all locations. It will do this by having spare capacity standing idle in case of breakdown and available to deliver to out of the way places on demand.

As an exercise, try applying these two concepts of 'efficiency' and 'effectiveness' to your favourite financial institution, school, public utility, transport system (including your car!) or health service. How efficient/effective is the human being?

The efficiency and effectiveness of the design processes will be compromises, as will the efficiency and effectiveness of the end result. A completely effective product will be expensive but it will be extremely reliable and robust. Military and life support systems need to be very effective. An optimally efficient system will cost as little as possible but it may go wrong every now and again, because the cost of replacing it is less than the probable cost of failure (at least, as far as the manufacturer is concerned). Domestic appliances need to be efficient in use.

Oddly, effective design leads to efficient products and efficient design leads (potentially, at least) to effective products. I'll explain.

Effective design leaves no stone unturned. Every aspect of a design is addressed with great attention to detail. The result is a design which is finely tuned to its prospective market and user environment, but so finely tuned, probably, that it might easily find itself out of its depth in other situations. Like a specialized species of animal, our incredibly efficient, low-cost product occupies a niche. A large niche perhaps, but a niche nevertheless.

Efficient design takes less time than effective design but may have holes in it because some aspects of the design will have been glossed over. These holes can be covered by over-specifying the actual end product, i.e. by making it better than it needs to be, to paper over any potential oversights. The effectiveness of the end product may not be intentional but is a by-product of the safety margin.

An effective product may find itself used in all kinds of unthought-of ways, particularly if flexibility of operation has been built-in as part of the safety margin. Unfortunately, innovative solutions tend to be missed by efficient design.

The microprocessor is effective. It does not occupy a niche; it occupies many niches, it is to be found everywhere. It is not efficient; a lot of its silicon may never be exercised by some software.

Think about which of these is the most important for your own design effort. Remember, the complexity of the ultimate solution does not always reflect the complexity of the actual design process which had to be undergone before the solution became apparent.

Some very simple (efficient) solutions may only arise after considerable (effective) thought has been given to the simplification. Brilliant simplifications of a design should be recognised for what they are and not dismissed as being worthless just because of a perceived lack of effort that went into their inception.

Useful activities and domains of concern

There are a number of activities which are highly appropriate to the large electronics company but which the amateur constructor would not be interested in using. For the individual, they are not efficient; in terms of the amount of time which they take to do, their potential usefulness is small.

There are also a number of activities which are employed by the individual and by the large company alike. For instance, all activities which are to result in an actual 'box of tricks' need a supplier of parts; these parts must be ordered and paid for.

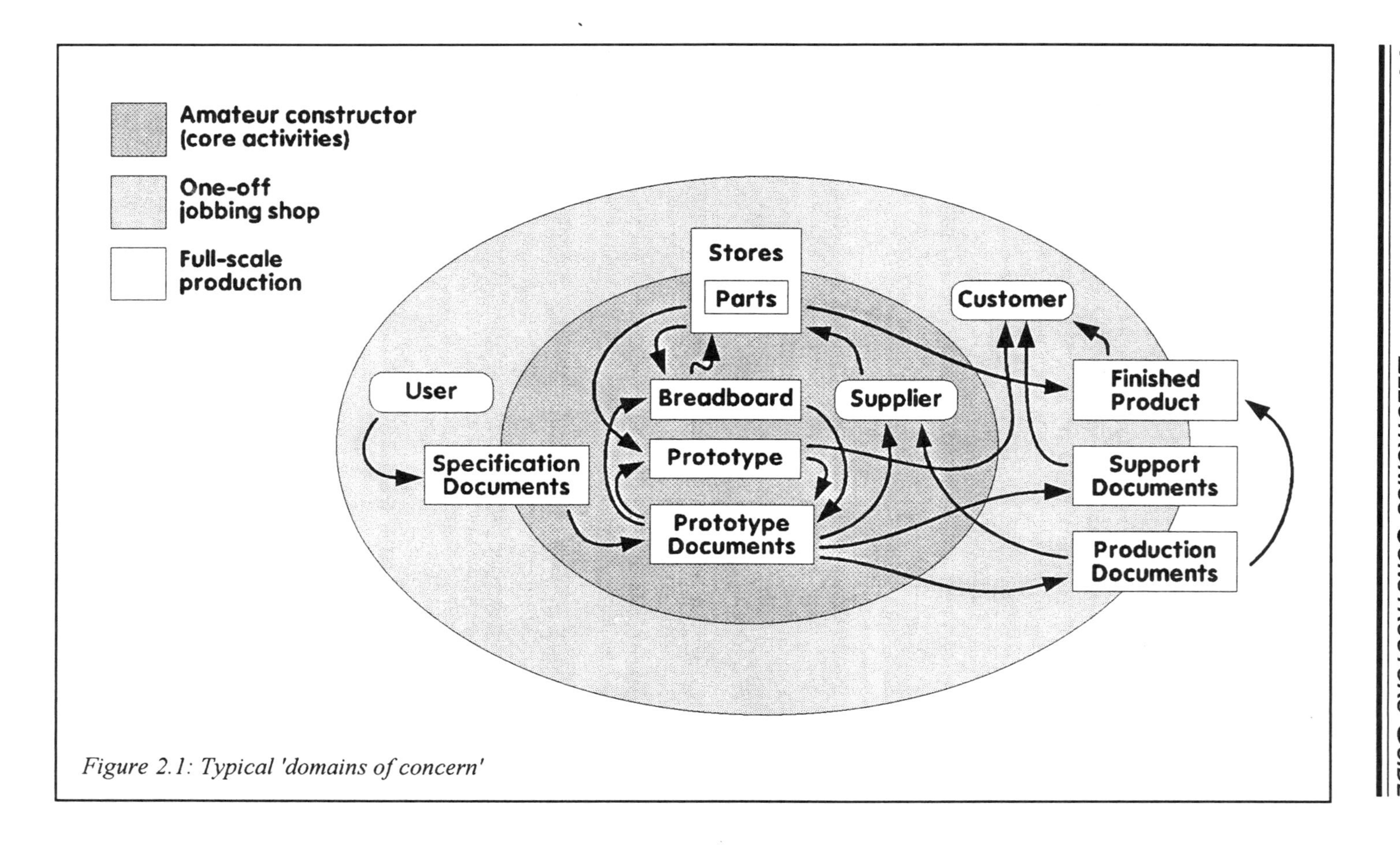

Figure 2.1: Typical 'domains of concern'

I have grouped these activities into 'domains of concern' in Figure 2.1, which shows, in a general way, how materials and information (in the form of documents) might flow around an organization involved in the design and supply of electronics products.

The shaded areas represent the domains of concern of the amateur constructor and the 'jobbing shop', the producer of one-off designs for particular customers. The jobbing shop domain encompasses and includes that of the amateur; to this extent, their concerns are the same.

The entire diagram is the domain of concern of any organisation involved in the full-scale production of its own designs. The design activities within the shaded areas may exist in well-defined cells within the larger organization. By 'full scale' we mean that the intention exists to produce more than one of a given design. Whether or not variants exist is not really important to our discussion at this stage.

We shall be looking more closely at the various kinds of documents in the next section. Here we will confine ourselves to looking at how they might be classified and to whom these general classes of documents are of interest.

All organizations of any shape and size will generate prototype documents of some kind, even if these are just a circuit diagram and a parts list for ordering. There will always be a supplier, as mentioned previously. Breadboarding will exist, even if it only happens in the form of 'test-as-we-go' on stripboard. There will be a parts store, however rudimentary, perhaps only existing as a plastic bag tucked away in a drawer.

Although perhaps an artificial distinction, I have defined the amateur constructor as having no user but himself and no customer but himself. If amateur constructors have actual customers or have had provided to them a specification from a user, then they are actually doing the job of a jobbing shop and I have classified them accordingly.

Specification documents, therefore, may be of passing interest to the amateur constructor. This is signified by the 'Specification Documents' box being only partly inside the inner shaded area. In most cases, though, well organised persons will have at their fingertips some form of document which gives an overview of how their design is supposed to behave, whether or not they intend to use it solely for themselves.

Support documents (user guide and so forth) are very rarely, if at all, produced by the amateur constructor. They may or may not be provided by the jobbing shop, as circumstances dictate. Organizations in full production will inevitably include such items as a user guide, as well as documents of a less technical nature such as registration and warranty forms.

In all this I have made the rather basic assumption that some concrete object is the outcome of the design exercise, i.e. there exists an intention, all things being equal, to build a prototype or engage in full-scale manufacture eventually. (This does not always happen.)

I have also used the word 'prototype' rather loosely in the case of the amateur constructor and the jobbing shop, where only one of a given design may ever be built. Strictly speaking, it implies that others of the kind may follow. For practical reasons I have used the word 'prototype' to mean 'one of a kind' as well as 'first of a kind'.

The arrows in the diagram signify the movement of information and materials from one activity to another. They can also be taken to signify 'dependency' (a lovely word which I have borrowed from software engineering). There are many opportunities for loops in the model as given. This allows for the possibility of iteration, the art of getting closer to a reasonably good solution in a step-wise fashion. If we are in a learning situation then iteration is likely to be more prevalent.

For instance, building a prototype depends upon having prototype documents to hand, typically a circuit diagram, an inter-wiring diagram and other such documents. On the other hand, the arrow heading in the other direction suggests that information flows from the prototype to the documentation.

This is indeed the case in many situations. Someone will hack away at the prototype; changes will be made and the documents modified (all in an orderly fashion, you understand). This happens around several iterations of the loop until an acceptable solution is found. Another example is in the circuit design itself; a first guess may specify AN Other op-amp as a general solution. Then, as we firm-up on component values and so forth, it may become apparent that offset nulling is needed, that impedances dictate that we use a particular kind of device or that savings can be made by using a dual or quad version.

Breadboarding or (if we are more confident of the ultimate success of the circuit) test-as-we-go on stripboard will generate further changes to be incorporated into the circuit diagram.

Seek to minimize the number of times you go round any loop by making as good a guess the first time around as you can. Always use the results of the previous iteration as a basis for deciding on any changes that are needed before the next.

One of the worst possible situations to be in is to not know whether or not you've succeeded yet. If you have no criteria for making a decision of that kind then it would seem that your specification is not complete (see next section). You must be able to test or measure what you've done and decide whether or not it is good enough.

Even in the sense of 'prototype' being 'one of a kind', the prototype will almost without question provide information to be incorporated into the documentation. A document derived after the fact is called an 'as-built'. It takes into account all the little things that no-one thought of until someone had a go at building it and found that they had to move *X* along a bit to fit *Y* in.

In the sense of 'prototype' as 'first of a kind', if we learn nothing from the prototype and do not incorporate this learning into the relevant documents, then why bother building it?

One of the distinguishing features of activities and information lying in the middle to outer reaches of Figure 2.1 is their concern with communication and record keeping. More people are involved in such activities and these inevitably become less of a personal thing. This effect is common to a lot of human activity (writing books is one of them!) To the lone wolf designer these are of passing interest, but they are absolutely essential to the running of any organization where co-operative work is the norm.

There is also a trend of scaling up as one moves to the outer layers, as well as a trend towards automation of some activities. These trends will differ in detail for different organizations due to the size and nature of the organization, its wealth and power and the attitudes of its members.

Usefully, these outlying activities can be brought into play as and when a former amateur constructor or jobbing shop makes a decision about going into production. If the decision to produce is completely retrospective, however, there will be substantial work to be done if the original idea has not been implemented with user-friendliness or ease of manufacture in mind from the outset. There may be constraints on finance and other matters, too.

Of course there may be organizations whose entire effort is focused on production of other people's designs. This model will not fit such organizations in detail. Nevertheless someone, somewhere, at some time, has built a prototype, drawn a circuit or produced a specification. The model is intended to give us an insight into planning and to suggest the means whereby we might plan the technical side of our own first product.

3: Documentation and the Design Model

Documentation is the first weapon in our armoury for planning a 'frontal assault' on a project. The soldering iron and snips wielding brigade might be put off by the word 'documentation'. They shouldn't be; documentation is just a four-dollar word for keeping track of things properly.

To be really effective, documents need to be neat enough and complete enough. By this we mean good enough for the job in hand at the time. Changes are inevitable and it is a thankless task making drawings up to 'presentation standard' before testing has shown which parts of the system can be finally documented. Presentation standard documents are best left to a tidying-up phase at the end of the design effort.

In this section we will look at documentation in general and see how it reflects the processes of design and production as carried out by our three kinds of organizations, the individual constructor, the jobbing shop and the full-scale producer. (You will be able to determine your individually precise needs for documentation by relating your project ideas to the material in further sections.)

Documentation can be broadly divided into temporary and permanent varieties. Temporary documentation, as its name implies, is destined for File 13, the bin, as soon as its limited useful life is over. There is room for the back-of-the-cigarette-packet drawing, even in carefully planned projects, but it needs to be kept firmly in the domain of temporary documentation, purely as a working scratch-pad for the owner. It must not become a means of communicating project details to other people.

Permanent documentation needs to be neat, complete and understandable. Not only will it be the means by which you record your efforts for posterity, but it could be that you will need to consult it yourself at some time in the future. It gives me a cold, prickly feeling when I come across a few poorly labelled sketches from years past, realising I'll have to make up a second unit using those scrappy notes . . . and I realise that I can't remember the component values for the hornswoggle filter.

General neatness apart, all wording on a document must be legible, even on a sketch. There is absolutely no point in scribbling down the value of a resistor which you've painstakingly calculated if you can't read the scribble the next time you come across it.

A quick word on annotation. I get confused between a '2' and a 'Z' so I always use a bar through the 'Z' to distinguish it. '5' and 'S' are sometimes difficult to tell apart; I often use small bars ('serifs') on the end of the S and make sure the top of the '5' is straight. Little 'feet' on a capital 'I' can usefully distinguish it from a '1'.

These are good habits and worth cultivating; they can save a lot of confusion. My use of a continental '7' with bar is a purely personal caprice. The continental '1' looks a little like an ordinary '7' - I remember my bank manager calling my attention to that once upon a time, back in the days when bank managers were sufficiently arrogant and I was sufficiently young.

All documents should have a title; a date might be handy. Unless you're operating in a situation where exact traceability is important, initialling and dating each revision is not particularly useful.

Let us take a moment to distinguish clearly between a drawing and a diagram, since we will be bandying these terms around a lot from now on.

A drawing is a sheet of paper. A diagram is a set of meaningful marks and notations. A drawing may contain one or more diagrams and there is no need, for small diagrams, for each to occupy its own drawing. Take the liberty of combining several related diagrams on one drawing if you want.

Incidentally, I am an A4 person. I leave the huge charts to people like architects who need them, since I find smaller sizes more convenient to print, more convenient to store and more convenient to transport. Split a large diagram into two logical sections on two separate drawings if you need to. It's convenient and painless and allows you to practise your partitioning skills too.

Documentation is like a kind of technical diary. Just like a diary, we can use it to record events or to plan events. As well as charting our progress in retrospect, good documentation makes

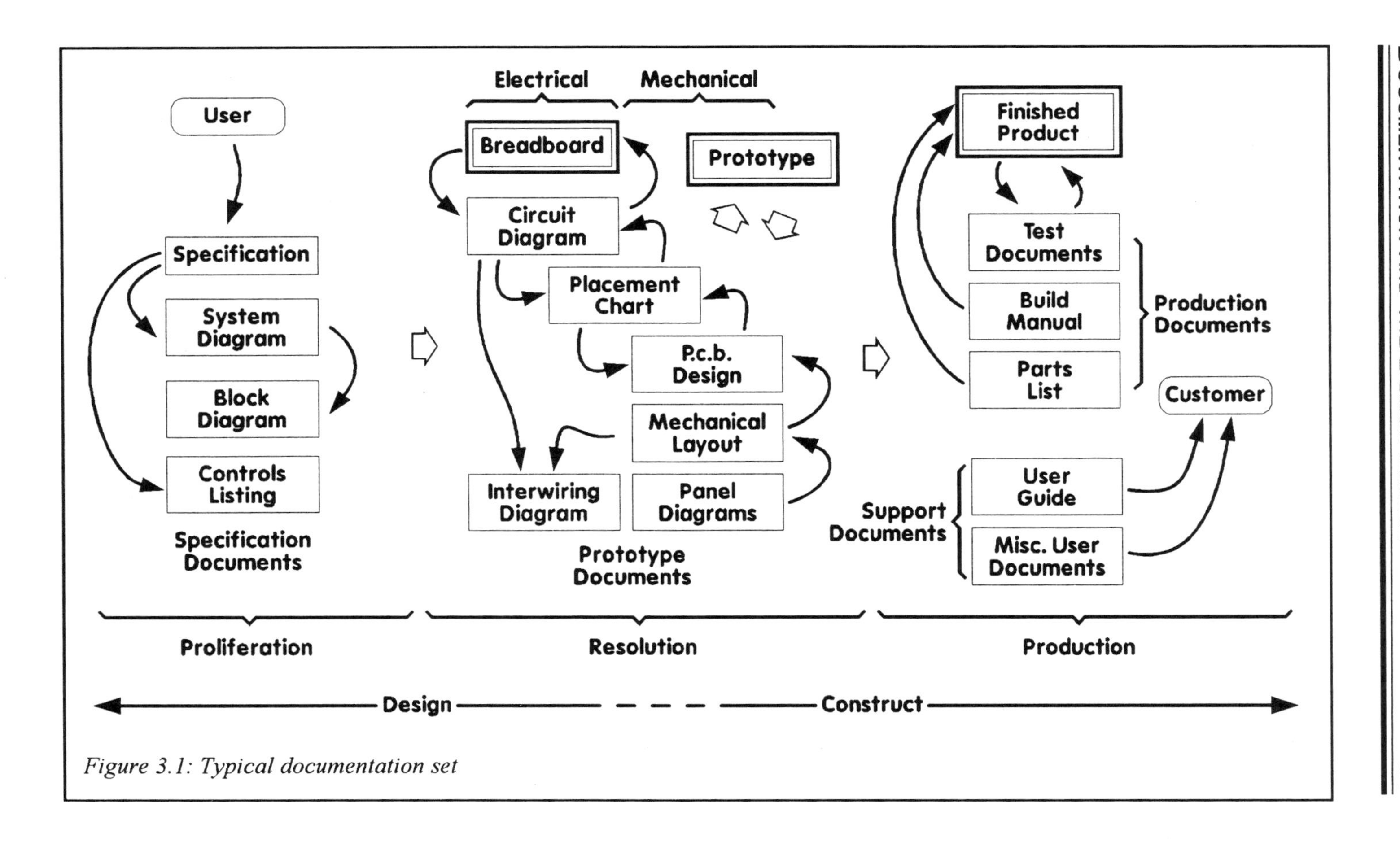

Figure 3.1: Typical documentation set

forward planning easy. Figure 3.1 shows some potentially useful kinds of documentation and their relationships to each other.

The general flow of work is from left to right. The plain boxes represent documents or sets of documents, whereas the double-walled boxes represent physical objects. The intention here is to show the flow of information from one item to its neighbours. The flow of materials is not shown on this diagram.

Arrows represent the flow of information. The large arrows represent broad flows of information from one major phase of the project to the next.

Broadly speaking, information is first acquired from the user. This is shown on the top left of the diagram. Even if this is not true in detail, there will be some user input of some kind; the end result of consultation is a specification. Information contained in the specification is then applied to the production of a controls list and a system diagram, from which a block diagram can be derived.

I have called this phase of the project the 'proliferation' phase since there is often a rapid generation of ideas at this time. This phase is also tied up with the idea of 'conceptual design'. We are not yet interested in the numbers side of things ('analytical design') although there may be an element of analytical design in the block diagram. Of course we are not totally disinterested in the numbers side; the specification itself may call for 20MHz operation, three channels, etc. We have just not progressed as far as the nitty-gritty calculation yet.

The second phase I have called the 'resolution' phase. Here we are moving from the realm of conceptual design into that of analytical design. Actual quantities and values gain in importance during this phase, with a great many calculations being made.

Physical objects (breadboard and prototype) make their appearance during the resolution phase; they are used to test the calculations of the analytical design and the information thus derived is fed back to the documentation and recorded. The breadboard is used to test the circuit design, whereas the prototype is used to test a number of aspects of the design, hence the broad arrows representing multiple connections between the prototype and the 'prototype documents'.

All pertinent features of the prototype must be recorded in the documentation, unless a positive decision has been made to abandon the project. Prototype documentation must be complete before entering the production phase.

There are two fairly independent entry points at the resolution phase - you could start with circuit diagrams on the electrical side of things or with the panel diagrams on the mechanical side. Working as an individual, alternating between these two aspects can provide interest and also allows you to transfer ideas from one side to the other in a less formal way.

Larger organizations could have two people or two teams each entering the resolution phase at the two entry points and looking at these two aspects separately. In this situation the communication between the two sides takes on a more formal nature and consists of meetings (productive ones we hope) and of sharing actual documentation (of the non-cigarette-packet type).

The electrical and mechanical aspects of the project meet at the level of the component placement chart, which details the physical locations of electrical circuit elements.

Once prototyping has been completed, the 'production' phase can be started. This does not mean that we start ordering equipment and components immediately. Timing is important and that implies a little planning.

Each phase of a project actually has a preliminary, a full-throttle and a winding-up 'sub-phase'. The preliminary sub-phase of production is: we need to prepare some form of guide for the production effort in terms of appropriate construction and quality standards and we need to allocate space and resources.

The two most important sub-phases are the production preliminaries and the winding-up of the resolution phase. There are a couple of good reasons for the particular sensitivity of these two.

First, a large amount of information is flowing from the resolution phase into the production phase. If such information is incomplete or incorrect then we are risking a waste of people's time which could have been prevented with a little forethought. This can have psychological consequences as well as financial ones. There is little which is more dispiriting than the continual change of 'plan' (if one could dignify it with that name) caused by inadequate preparation.

Second, we are probably spending (risking) large sums of money, not only in terms of stock incorrectly ordered or lying unused over time, but also potentially in terms of salaries (due to the wastage of time already spoken of) or due to having production plant and machinery, bought and paid for, lying idle.

These considerations are all the more important in larger organizations where the potential for waste is even greater. Prepare properly; a bad first move is hard to recover from. As with all things, errors will creep in. Mistakes are waiting to happen. But make mistakes small, make them early on and most important, do not propagate them from one phase to another.

Figures 2.1 and 3.1 are related. Figure 3.1 is, from the point of view of documentation, an expanded version of Figure 2.1. Comparing Figures 2.1 and 3.1, you may have spotted that the amateur constructor is mostly concerned with things at the resolution phase, the other two phases existing perhaps in rudimentary form. Generally, the jobbing shop will need to further concern itself with the proliferation phase and the factory will need to concern itself with all three phases by including the production phase too.

Neither of the models presented in Figures 2.1 or 3.1 are to be taken wholesale as the answer to all design problems. Apply them intelligently, modifying them for your own use and removing parts or adding your own parts as you see fit.

By ensuring proper completion of one phase before embarking on the next, we can plan more effectively and reduce the amount of to-ing and fro-ing between project phases. Thus we minimize the amount of information which needs to be transferred between phases and remove potential confusion traps.

To avoid cluttering up Figure 3.1, I have not mentioned feedback. Feedback can take place as a kind of backward leakage against the direction of the large arrows. The predominant flow of information is still from left to right, though; the agenda for any phase has been set in the preceding phase.

If you're running a jobbing shop then you're into feedback in a big way and everything you do is potentially up for change; you have, after all, supplied the customer with the only prototype.

If you're building a vast number of identical units then feedback of various kinds will be important. But feedback needs to be controlled. On no account must changes to the specification be made, for example, on the strength of the preferences of the person preparing the user guide. On the other hand, do listen to production people tipping you off as to ways of simplifying manufacture, where these do not affect the performance or appearance of the product. Make your groundwork solid. Strenuously avoid backtracking, in any significant way, from any given project phase to its predecessor.

As well as talking to production about production-oriented matters, a further kind of feedback might be provided by customers coming back (as 'users' within the meaning of Figure 3.1) and influencing the specification of proposed new models.

By being organized in the ways I have outlined above you will lose in the short term but gain in the long term. Someone who is not well organized may look as though they are working hard since the bench or desk or indeed the entire premises is quickly littered with all kinds of drawings and half finished bits and pieces. Well, maybe they *are* working hard - but working smart is better.

A really well-organized person may seem to be exerting no effort at all. The thing is, if everything falls neatly to hand when needed, then there is no fuss and no frustration. Measure yourself and other people by results, not fuss, with due credit for a good piece of documentation, which is a result in its own right.

Keep both your documentation and your head straight by using this model or a similar one of your own. You will not avoid mistakes completely that way but they will be easier to spot and easier to put right. You will have finished your project faster and it will stand up to closer scrutiny in the end. The same comments apply to other models which I shall be describing for system partitioning and for printed circuit design.

Treat any model as a kind of check-list if you want. The whole idea is to lessen the time it takes to cycle round the design, without compromising the quality of the design or omitting something which is going to cause a panic at the last minute.

Remember the old saying: models are to be used, not believed. We are not asking for an article of faith here.

Individual sections within this chapter look at each individual document, with examples where needed, to enable you to decide whether such a document is appropriate to your efforts. It also offers a guide as to how to produce such documents in the most painless way, and what information each should contain.

Specification

The specification is the primary document which should act as a guide for all subsequent activity. A simple specification might only occupy one A4 sheet of paper, only take an hour or two to put together and properly done could potentially save days of fiddling and indecision. It may also conveniently form the basis for an agreement between two parties where the designer and eventual user are not one and the same.

It should not specify how something is to be done or implemented, rather it should specify *behaviour*. For example, statements such as 'the circuit shall function up to 2MHz' are quite in order, as there is then obviously a need to function at this frequency.

On the other hand, saying 'the transistor BC107 shall be used' is a waste of time and places an unnecessary constraint upon the designer. Maybe you happen to have five million BC107s stashed in a warehouse somewhere. In this case, recognise that this is a commercial constraint and not an engineering constraint.

Occasionally the presence or absence of some piece of technology, a microprocessor say, is regarded as a selling point, yet it possesses no engineering value or advantage to the user. When faced with this situation, try and make engineering-wise decisions.

Avoid using the terms 'optimize', 'minimize' and 'maximize' in a specification. It is not always easy to tell whether or not something is optimum. Also, such activities are thoroughly open-ended and could potentially carry on until the designer is collecting her pension. Instead, quote an actual figure which is the acceptable limit or constraint of some performance criterion.

At the same time, allow the designer credit for some intelligence; terms such as 'hand held' imply battery power and not, in all probability, a 12-volt lead-acid car battery.

The things which a specification must contain as a minimum are a list of the functions which the device must perform and, where the device interfaces to something else, the name and nature of the other system and the exact nature of the information transferred.

The other things the specification might also usefully note are: power supply constraints; environmental constraints (e.g. temperature); size and mass constraints; for measuring instrumentation, allowable tolerances on measurements; interference compatibility requirements; and the names of the controls which activate or modify the functioning of the device. This list is not exhaustive. Above all, think about how the resulting system is to be used!

One of the few points at which a specification might touch upon actual components is when specifying the nature or position of controls. It is quite legitimate to mention these since they may strongly influence the use or misuse of the equipment. This is particularly important in marine use or in avionics or where a simulator or plant control cabinet is being designed. Good panel layout is discussed further in a later section.

If you're really stuck, mention every knob, display and button and every external connector (in terms of behaviour, remember). If you go on to include weight, size and preferred materials/paint job then you have probably just about cracked it.

In those situations where you are required to produce a specification for someone else, get a copy to them as soon as it's been prepared. Unless you know them well enough to gauge their reactions, do no more to the project until they've given you approval to continue.

It's important to think about the end of a project. The specification should say what is needed; once these needs have been satisfied, the project is finished. Remember that your time is valuable and resist the temptation to add bells and whistles at a late stage of the design. All bells and whistles should have been thought up before embarking on any resolution work.

Incremental specification, the art of moving in well-defined steps, learning from the experiences of the previous stages of the project, is a useful thing to do for larger projects. Make an agreement with the end user on a series of points at which you can usefully review progress. Use these break points to take stock, revise future plans and tidy up existing documents. Use them as a basis for a payment schedule too.

It is a great temptation to work without a specification or ignore it but the potential for disaster, especially for fairly large projects, is too high. I remember well working without a proper specification once. A year after I'd finished my part, I was still making noises about being paid. Eventually, it turned out that the firm had not used all of the work I'd done and would I mind please not being paid? Of course, I had no real proof of having done anything constructive at all. I was lucky to get away with 50% of the bill. The solution: go and work for someone else who values your time.

A specification is still useful to an amateur constructor building for his own use, although of course it need not be so comprehensive as that required commercially. It will certainly not be cast in the role of a legal document. If done properly it will still provide a good working guide to keep the design effort on track.

If at any subsequent time it seems that there is a requirement which is realistically impossible to fulfil, then it's time to go back to the specification and see about relaxing it in some way. It's a wise move to scan a specification and choose to experiment with the difficult parts first as this can short-cut any problems. Having intractable problems rear their heads at a late stage is a hazardous event as it could mean a lot of reworking.

Only experience will tell you where these difficult parts are likely to be, but usually they will involve learning curves. Parts of the system which use unfamiliar chips and uncommon circuit configurations are good candidates. Looking at the block diagrams might also throw up some subsystems which will repay early investigation.

Another good point is to get other people, those who are nominally responsible for some other part of the system to which yours interfaces, to commit themselves to a working plan. If they change their minds at some time in future then it will invalidate all your efforts.

Usually, the need to change the specification becomes apparent during circuit design and testing or while trying to fit the contents of the system into the space available. As an example, consider the case where we cannot fit a certain switch to the enclosure since it would then foul the p.c.b. inside. We have discovered this unfortunate clash while quickly sketching out the insides of the cabinet for our 'first approximation' mechanical layout.

Ouch! OK, but we've caught it early on, before committing ourselves to an actual wiring scheme, circuitry or p.c.b. design. We have a number of options at this point. Re-working the specification, after due discussion, will give us either a new enclosure type or a new switch. A better alternative is to look hard at the physical layout and decide that we can mount the p.c.b. in a different place.

Better yet, quickly scan the literature for a different switch which has the same appearance and behaviour as the originally specified switch. That could be a time-saver.

Either way, we have avoided getting as far as the production phase before discovering the error. We are re-working on paper rather than re-working expensively on the system itself. The costs of re-working will be mentioned again under the heading of testing. If you wish, we are using a 'paper test' or 'paper experiment'. Such experiments are quick and cheap to do.

Here follows an example specification. I knocked it up in an hour or so. Although there may be some inconsistencies, it would still form a good basis for further development and discussion.

```
Data Buffer type R10

General:

The unit is to be a buffer to temporarily store
serial data transmitted from a computer to a
serial device such as a plotter. The buffering
capacity shall be 256k bytes expandable in 256k
byte increments to a maximum of 1M byte. The
buffer shall receive data until the buffer
becomes full. At this time, a 'FULL' LED shall
light and further reception of data shall be
disabled. When, due to onward transmission of
data, 10k of memory is freed, the 'FULL' LED
shall go out and further received data can then
be accepted.

Handshaking shall be by both hardware and by
XON/XOFF protocols, selectable at the rear.
Available baud rates are 300, 1200, 2400, 4800
and 9600. 7 or 8 bit working is selectable.
```

There is one stop bit and no parity.

The unit is to be mains powered.

Controls and displays:

a) Red LED which is on whenever the buffer has halted transmission. Labelled FULL;
b) Green LED which is on whenever power is applied. Labelled POWER;
c) Yellow LED which is on when the buffer is empty. This LED blinks whenever data is being received from the computer. Labelled EMPTY;
d) Switches accessible from the rear for choosing from the set of available baud rates, protocols and bit settings;
e) Power connector at rear, IEC standard 3 pin chassis plug;
f) Power switch at rear, rocker type;
g) 9 pin Dee x 2 at rear for serial receive and transmit (PC compatible pin-out).

Implementation:

If microprocessor control is decided upon, the preferred type is the Z80.

Enclosure:

In order to ensure adherence to the house style, a box from Acme Plastic Box Co.'s Buzz range, in a buff colour, should be chosen to blend in with our existing modem products.

I've deliberately included a couple of borderline cases in this specification.

Firstly, there is the inclusion of advice on choice of microprocessor. My comments have said that one should not enforce a choice of component on the designer. In fact microprocessors require support, in the form of software, and the organization might already have invested in software development systems. Thus, there may be a commercial constraint on the choice of microprocessor.

All things being equal, i.e. *where such a constraint does not compromise the performance of the product*, the preferred chip, microprocessor or whatever should be used. Where there is some difficulty in making the preferred component do the job, you will need to weigh up the costs of persisting with a less than ideal chip and the costs of re-training and re-equipping with the development systems for a new type.

The other woolly statement might be in the choice of enclosure. We are specifying an appearance, however, which is a kind of behaviour, so this kind of constraint is fair game. Be prepared to adjust, though, if all the goodies won't fit into such a box.

Note that in this case there is an existing product range; the designer of new kit to be added to this range should take some cues from the existing products, especially in terms of the general appearance, the labelling and the arrangement of controls and connectors.

Controls list

A list of controls and displays is a good idea, unless it already forms a detailed part of the specification. Other kinds of 'furniture', as it is known, can be added to the controls list or perhaps kept on a separate list. The most important of these additional items of furniture are connectors.

Don't miss out any mains connectors or battery hatches which are implied by the power supply specification. Also don't forget to mention any flying leads or other bits and pieces which must also be accommodated somehow. These may not always be mentioned explicitly (our example specification above is quite comprehensive in this respect).

What you probably won't know yet is whether any heat sinking is needed. If you're unsure then just jot down a note to remind you to look into that side of things once the design has advanced to the point where cooling requirements have been worked out.

Perhaps the enclosure that you're thinking of using has space reserved for units requiring cooling.

Table 3.1 shows a typical controls and furniture list with columns for control labelling, type, part number (for easy relation to a manufacturer's catalogue) and last but not least, a column for mechanical data (fixing centres, clearances and so forth). It represents part of the controls for a hypothetical laboratory filter. It is just a fragment, but will hopefully convey the idea.

Label	**Positions**	**Type**	**Part No**	**Mechanical**	**Notes...**
Power	ON/OFF	Rocker	129-002	20x40 hole	
Frequency		Group label			for Coarse & Fine
Coarse	10, 1k, 100kHz	Rotary 3w?p	129-030	12mm hole 23mm body	
Fine	0-99	10t pot	130-209	12mm hole	
		Heatsink			

Table 3.1: Part of a typical controls/furniture list

Usually, you will derive the information in the first three columns from the specification itself. Then use your judgement (some tips are given in a later section) to choose an appropriate component type which matches the specification.

Again, I've chosen a slightly awkward example, which includes a coarse and a fine frequency control. Often a single label and a box will enclose two or more controls with subsidiary labelling. For completeness I've included the 'group box' as a control (if you're a Windows programmer you may already be familiar with the idea of 'groups' as controls).

For obvious reasons, a group box has no part number and no fixing holes. But it does take up space on the panel and needs to be thought about.

Glean the remaining information from the supplier's catalogue you're using and enter it in the relevant columns. You won't need to go repeatedly hunting around the book if all this information is to hand. Note the catalogue and page number too if you like; it's all useful stuff.

You would not need a list of furniture, by the way, if the project were simply a plug-in board, although there are some instances, even with this, where some kind of p.c.b. mounting furniture needs to accommodated or fixed to the panels of a rack system.

System diagram

Block and system diagrams are an excellent guide to the potential complexity of a project. In some ways the distinction between system and block diagrams is a little artificial, in that there is a kind of spectrum or hierarchy of diagrams from the 'big picture' system diagram down to the nitty-gritty detail of the circuit diagram itself. However, it is still a useful distinction to make.

System diagrams are more useful for large efforts, typically where electronics is sited in several cabinets, often remote from each other.

They are useful to get an indication of how much wiring is needed between subsystems and cabinets. For those projects which can use them, they represent the first stage of partitioning.

A typical system diagram is shown in Figure 3.2.

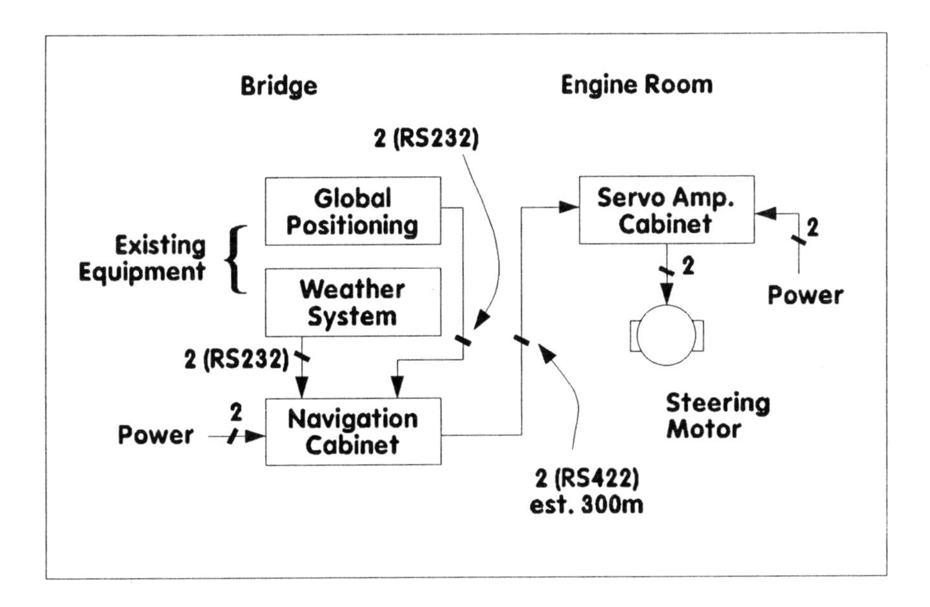

Figure 3.2: A system diagram

I've been a little imaginative with this diagram. I don't know how ships work to this level of detail, but I believe that my guesses are plausible and workable. The ability to guess plausibly (backed up by subsequent analysis, of course!) is a boon. For the purposes of Figure 3.2, I've imagined that we need a pilotless ship which can steer itself around the globe, along a number of tracks from ports on one continent to ports on another, and that we need to supply a steering system to do this.

There are two main cabinets in the system, the navigation cabinet and the servo amplifier cabinet. Existing equipment consists of a global positioning system (GPS) and a weather system. I've guessed that the GPS and weather systems will talk to the navigation cabinet via a serial link and I've also guessed that the navigation cabinet houses the 'intelligence' which makes decisions about steering.

Any such system diagram is a first guess and needs to be explored more thoroughly by producing block diagrams for the contents of each of the cabinets.

Guessing about the proximity (or otherwise) of one cabinet to another has enabled me to make guesses about the kind of serial transmission which we need, RS232 for the local communication and RS422 (differential) for the long-haul stuff from the bridge to the engine room. These guesses are based on experience. Perhaps I also read the GPS manuals and obtained the information from there; reading manuals isn't cheating!

As a result of drawing a diagram like this, it becomes a lot easier to think about the kinds and quantities of connectors we need for each cabinet. The system diagram carries information relating to the number of wires needed in any given cable. The thick strokes through the connections signify that these are cables and not single wires; the number alongside each stroke is the number of cores used.

The specification and furniture list will provide clues to cabling requirements. You should also gain some impression of the amount of cabling needed by studying the equipment to which your unit needs to be connected.

You either have a free hand in choosing cable and connectors, or you do not. In this instance, reading the manuals for the GPS, the weather system, the ship's power supply and any existing steering motors will provide lots of useful information on interconnection.

Where restraints are imposed by existing equipment, use them as design cues or design opportunities.

On the other hand, you have a free choice of cabling and connectors between the navigation and servo amplifier cabinets, given that these comply with any regulations concerning shipboard use.

Where actual connector types are not firmly specified, use your thoughts about the system diagram to bring the controls list up to scratch, once you have made any decisions about connectors.

In this form, with cabling information in it, the system diagram is not only a good planning document, allowing generation of the block diagrams, but also becomes useful as a part of the installation guide, commissioning guide or 'build manual' produced during the last phase of a project.

Notice what has been left out of the system diagram; there are no voltages, currents, communications protocols or what have you; we probably don't have some of that information yet, anyway.

Block diagrams

A block diagram is more useful to describe the contents of a single subsystem, cabinet, board, etc. The system diagram could be thought of as a kind of 'super block diagram'. Most projects, however small, will benefit from use of a block diagram of some kind.

To arrive at the system and block diagrams we need to partition. As far as the system diagram is concerned, partitioning might have been done at the specification stage. So much the better. The block diagram, however, is our halfway house to a real circuit and will rarely be partitioned at the specification stage. The actual techniques of partitioning (or segmentation) are described at length in Chapter 5.

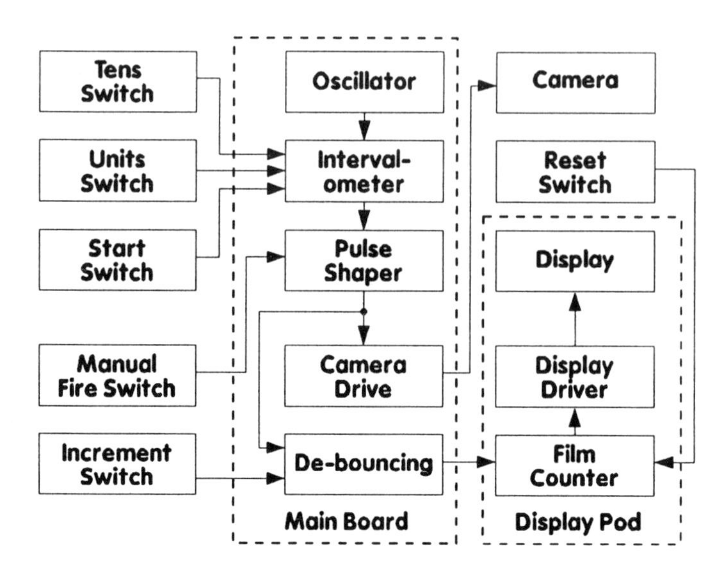

Figure 3.3: A block diagram

Figure 3.3 shows a typical block diagram.

This diagram, incidentally, is derived from a camera control system which I built and which is intended to fire a photographic camera at regular intervals of time.

Note that each block has a name and is connected to at least one other block.

Each block of a block diagram often (but not always) translates, at the circuit level, to a one or two i.c.s and their associated small components. In this case, it turned out that the oscillator used one-and-a-half chips, the debouncing circuit using the remaining half. The film counter and pulse shaper each used one i.c. and the intervalometer and display driver two apiece. The camera drive is just a relay.

I have not included a block for the power supply on this diagram. The reason for this is that the power supply is connected to 'everywhere'; it would just confuse the diagram to attach a general purpose power supply, by numerous lines, to everything else.

Sometimes the power supply is included on the diagram but not connected up, i.e. the connections are assumed; this can be a useful reminder to include it in the circuit diagrams. Special supplies, connected to just one other place, need to be shown as being so connected.

Note the use of arrows on the connecting lines in the block diagram. The system diagram doesn't use these as it is cable oriented rather than wire oriented. Information might flow both ways simultaneously through a cable (think about serial communications, for example) so an arrow on a system diagram is probably ambiguous or, at best, uninformative.

As it is, an interwiring diagram (see below) usually shows the direction of information flow for each individual wire in an unambiguous way.

Lines on a block diagram need not translate to individual wires or physical connections. They merely represent the direction of flow of power or information. Take the 'tens switch' of Figure 3.3, for example. Until we get down to the circuit level we really don't know how many wires there are between it and the main board.

The specification might say that this switch is a rotary ten way switch and we might then guess that eleven wires were used or, if we could work in BCD (binary coded decimal), five wires (plausible; actually, ten were needed, as it turned out).

Again, the interwiring diagram is the best place to record this information, with possible duplication on the actual circuit diagram.

Panel diagrams

By this time, you should have a list of all the things which stick out of the box, so to speak, in the form of a controls/furniture list. As a result of this activity, and as a result of deriving the system diagram, where that is useful, you will have developed some notion of what the system is going to look like physically.

With our attention now focused on panel diagrams (or circuit diagrams if we choose that entry point) we are firmly engaged in the 'resolution' phase of the project.

Each cabinet, box or enclosure in the system will generate a panel diagram for each face on which panel furniture is to be fitted. Typically, there will be two panel diagrams (front and rear), the theoretical possible maximum being six, I suppose.

Where a small 'inset panel', let into a much larger panel, carries panel furniture, draw the small panel to a conveniently large scale and its actual location within the bigger panel to a smaller scale.

In making up panel diagrams, you will use the controls list, or the specification where this contains an adequate list itself. The actual design of panels and the decisions which need to be made about them are covered in detail later.

Figure 3.4 shows a design for a front and a rear panel.

Note that these panels are drawn as they might actually appear. In this form, they are suitable for showing to the users for their comments, and as such are probably the last contact you will need to have with users until the actual unit is built. Draw to scale where possible, including the external furniture in the form of knobs and so on.

These kinds of diagrams give you a feel for how the unit will look once finished and help you organize your thoughts about what is to happen inside the box. They also provide a check on whether or not you can really accommodate the controls on the panel, along with all the panel markings required. (Controls may still foul each other behind the panel - see below.) It's a dreadful sinking feeling you get when you've committed yourself to printed circuits and other mechanical parts of the design and then discover that your chosen enclosure's panels do not, after all, have room for all the panel furniture.

When you come to preparing working diagrams for others to work from, you will need to split diagrams such as those of Figure 3.4 into two, discarding the external fixtures (knobs, levers) in the process. One part will show panel markings only and is suitable for the printing works or whoever else is to mark the panel up. The other shows all the mechanical details, i.e. hole locations and diameters and so forth and can be given to the machine shop. This latter kind of diagram, properly speaking, forms part of the mechanical layout set.

The actual arts of successful panel design are discussed in detail in a later chapter.

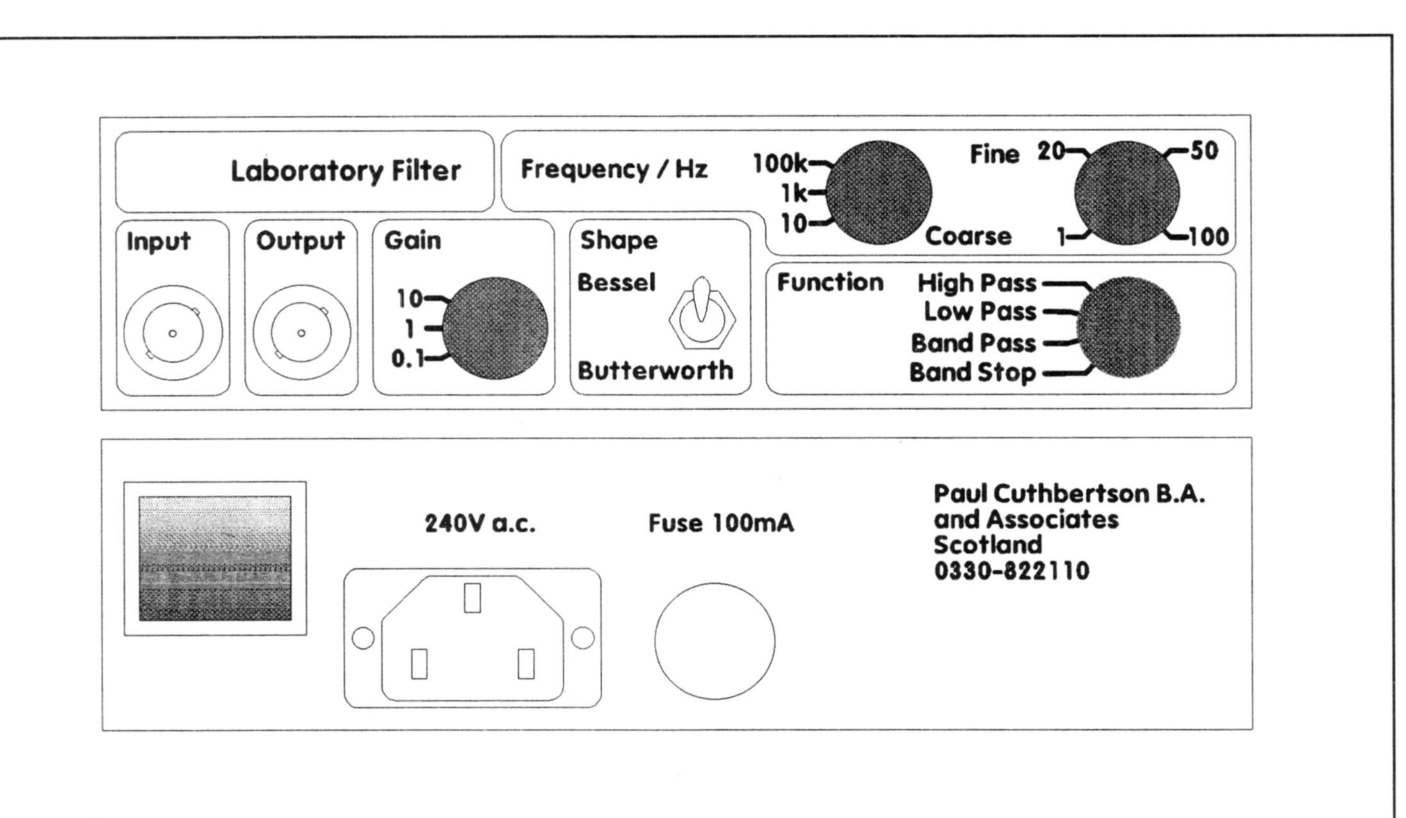

Figure 3.4: Panel diagrams

Mechanical layout diagram

Where the panel diagrams give us a useful two-dimensional picture, mechanical layouts give us a 'ground plan' and one or more elevations, which allow us to fit large components and check for interference between boards and panel furniture. One thing the panel diagrams do not address is depth; we must use other diagrams to represent this aspect.

Where interference is detected, try to alter the internal layout of the unit in preference to altering the positioning of the controls.

Draw in the maximum extent of any controls both in front of and behind the panel. In that way, any interference can be found right away. Working to full size on small panels gives a better feel than using a small scale.

At this stage, we're guessing about the sizes of boards. There is no harm in assuming, for the moment, that a board will occupy a certain position, especially if card mounting bosses or rails already exist. It's not a bad idea to note p.c.b. size constraints on that sheet of paper destined to become the p.c.b. sketch, purely as a reminder to take size and fixing constraints into account as soon as you've made any tentative decisions about this.

In fact, it may be useful to have a blank piece of paper, or a paper with just a title on it, ready to hand to carry these comments as information is transferred between one aspect of the design and another. When you eventually get to the p.c.b. design, for example, you will have a little list of comments, generated during the course of thinking about other things, which will give you a good kick-off into the p.c.b. design itself.

Where the system diagram does not exist, that is in those cases where a single enclosure is being used, there will be only one set of mechanical diagrams.

For most electronics and any available enclosure there's often a great mismatch between the space available to fit the actual electronics inside and the space available to fit the controls on the panels. Either there are only a few controls and masses of electronics resulting in a rather bare looking front panel, or there are lots of bells and whistles and a huge empty space inside the box.

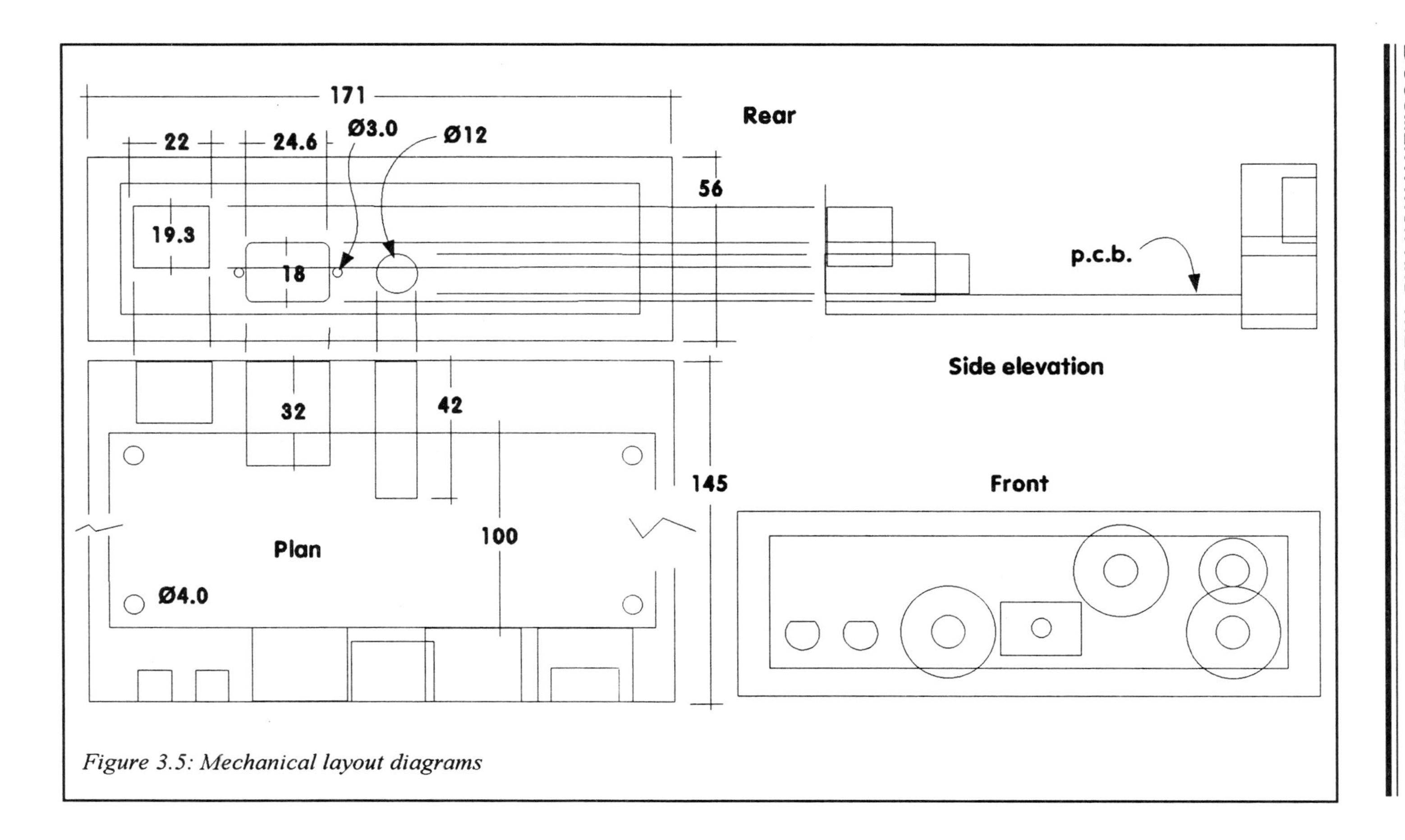

Figure 3.5: Mechanical layout diagrams

One common mistake, where controls are few, is to use an enclosure that's too small. There is a real art to getting the balance right, but proper use of the various physical layout diagrams will catch any problems as they arise.

Although not shown on Figure 3.1, the p.c.b. design can influence the mechanical layout in a small way, in that, if it found that the components will not all fit on the board, then two boards may have to be used. This may or may not impact on the mechanical design aspects and choice of enclosure.

Figure 3.5 shows a typical set of mechanical layout diagrams. In fact, these are for the 'Laboratory filter' whose proposed panel layouts are shown in Figure 3.4.

Sizes must be entered onto a mechanical diagram. It's a good idea if it's drawn properly to scale, too, since any interference will be more easily spotted. In this particular example, which is a bad example, there is quite a bit if interference; so much so, in fact, that it would be worth thinking about using the next size of enclosure. If you look closely you will see that the mains connector fouls the p.c.b., a rotary switch fouls the chassis, another switch fouls the chassis and the potentiometer and both of these switches potentially foul the p.c.b. too.

Fortunately, we did it all on paper - so it only cost us an hour's thinking time. Thank goodness!

Interwiring diagrams

It's time to introduce a valuable tool for keeping wiring in order. The interwiring document allows us to specify precisely which wires go where, how big they are, what colour they are and how they are to be identified.

It might seem that we are approaching this thing backwards. After all, does the circuit itself not dictate the wiring?

Well, actually: no!

In fact, wiring dictates the circuit. We should have no more wires than is necessary to carry information (or power) between the various items in the enclosure or to the front panel. Unless experience dictates that one signal should be carried by more than one wire, or that several signals be carried by a single wire, than the usual rule is one signal, one wire. All of these issues are

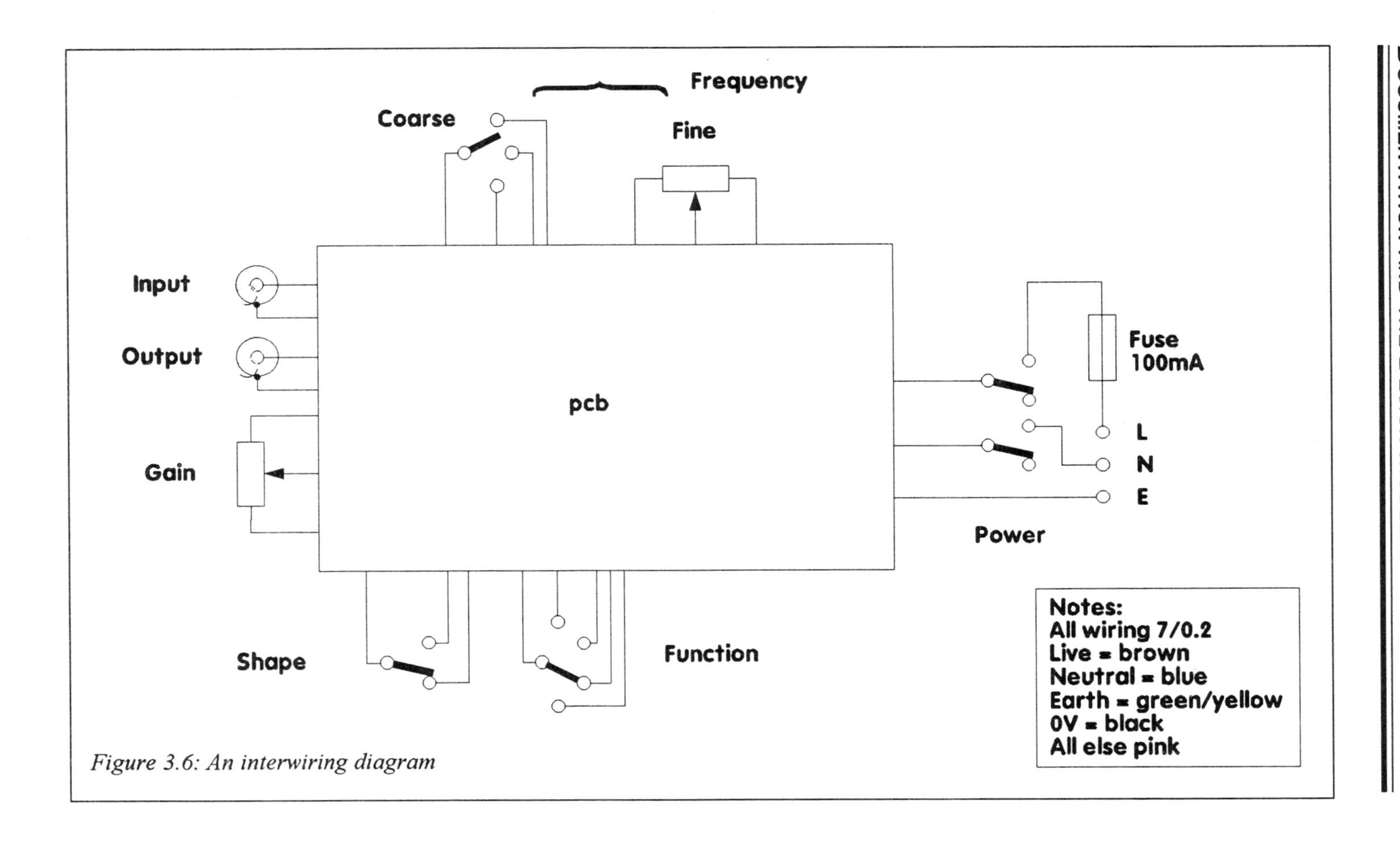

Figure 3.6: An interwiring diagram

No.	Colour	Gauge	From	To	Group	Signal	Comments	Off
25	Pink	7/0.2	J1f/1	J3m/1	A'log	Vref	Stirrer Voltage Ref	
26	Pink	7/0.2	J3f/1	VR1/cw	A'log	Vref	Stirrer Voltage Ref	
27	Black	7/0.2	J2f/1	J3m/2	A'log	Return	Stirrer Zero Volts	
28	Black	7/0.2	J3f/2	VR1/ccw	A'log	Return	Stirrer Zero Volts	
29	Yellow	7/0.2	J1f/2	J3m/3	A'log	Signal	Stirrer Wiper	
30	Yellow	7/0.2	J3f/3	VR1/wpr	A'log	Signal	Stirrer Wiper	
31	Pink	7/0.2	J1f/3	J3m/4	A'log	Vref	Hopper vane Voltage Ref	
32	Pink	7/0.2	J3f/4	VR2/cw	A'log	Vref	Hopper vane Voltage Ref	
33	Black	7/0.2	J2f/2	J3m/5	A'log	Return	Hopper vane Zero Volts	
34	Black	7/0.2	J3f/5	VR2/ccw	A'log	Return	Hopper vane Zero Volts	
35	Yellow	7/0.2	J1f/4	J3m/6	A'log	Signal	Hopper vane Wiper	
36	Yellow	7/0.2	J3f/6	VR2/wpr	A'log	Signal	Hopper vane Wiper	
37	Pink	7/0.2	J1f/5	J3m/7	A'log	Vref	Hopper pump Voltage Ref	
38	Pink	7/0.2	J3f/7	VR3/cw	A'log	Vref	Hopper pump Voltage Ref	
39	Black	7/0.2	J2f/3	J3m/8	A'log	Return	Hopper pump Zero Volts	
40	Black	7/0.2	J3f/8	VR3/ccw	A'log	Return	Hopper pump Zero Volts	
41	Yellow	7/0.2	J1f/6	J3m/9	A'log	Signal	Hopper pump Wiper	
42	Yellow	7/0.2	J3f/9	VR3/wpr	A'log	Signal	Hopper pump Wiper	
43	Pink	7/0.2	J1f/7	J3m/10	A'log	Vref	Cooling water Voltage Ref	
44	Pink	7/0.2	J3f/10	VR4/cw	A'log	Vref	Cooling water Voltage Ref	
45	Black	7/0.2	J2f/4	J3m/11	A'log	Return	Cooling water Zero Volts	
46	Black	7/0.2	J3f/11	VR4/ccw	A'log	Return	Cooling water Zero Volts	
47	Yellow	7/0.2	J1f/8	J3m/12	A'log	Signal	Cooling water Wiper	

Table 3.2: A list type wiring document

addressed in detail in a later section. For now, take a look at Figure 3.6 and Table 3.2, which show two variations of the interwiring document.

I've called these interwiring 'documents' since there are two approaches, an actual diagram or a list. For modest quantities of wires, say up to a couple of dozen wires or so, a diagram is fine. For larger quantities of wires, a list is a better bet. For one thing, a long list of wires can be conveniently manipulated by computer; sorting is easier and there is always the possibility of being able to estimate quantities automatically.

Figure 3.6 is an interwiring diagram for the hypothetical Laboratory Filter. I have guessed about the switch contact configurations, choosing the simplest possible which could do the job. I've also guessed that a single pot will do for the fine frequency control - perhaps a dangerous assumption! Once I've got as far as a tested breadboard and circuit diagram there may be details I need to amend in this diagram.

Table 3.2 is part of a wiring list for a hypothetical control box. In this kind of situation, wires need to be numbered since there are so many of them that life would be made difficult without numbering. The columns for wire number, colour and gauge carry fairly obvious information.

The next two columns say where the wire comes from and where it goes to. J1f/1, for instance, means pin 1 of connector J1 (female side). You will have decided already about the probable nature of the required connectors. You will also have decided which need to be bulkhead or panel mounted and which are on the end of an umbilical cable ('free', as the saying goes).

Be specific and give connector and pin numbers. A good wiring list is a design tool, not a record after the fact. Be prepared for surprises; in large systems, a decision to use connector type 'X' may very well result in several of connector 'X' and the physical plan may need amendment accordingly, as the number of wires swells. Without producing a proper wiring list, surprises of that kind are common and it can be awkward trying to fit more in after you've committed to the mechanics. On the example given, each control potentiometer uses six wires. It could have been more.

Remember that each row of such a list corresponds to a single piece of wire, not a whole route. Thus, Stirrer Voltage Ref. at the top of Table 3.2 must traverse two pieces of wire to get from pin 1 of J1 to the clockwise (cw) end of VR1 (the stirrer control).

If you're doing this on a computer, copy down a bunch of things and then edit them, saving time. For instance, you've 20 micro-switches in your new robot arm and they're all wired with similarly coloured wire. Get one fully described and then make another 19 copies of it, editing connector pin numbers, etc. to suit. Don't copy the wire number column! This saves time and typing but you must keep track of where you are so as not to duplicate by forgetting to edit after the copy.

I tend to group wiring in a list like this according to the equipment it is feeding, as opposed to sorting the list into kinds of wire, connectors, colours and so forth. If done this way with wire numbers granted as soon as the wire is entered onto the list, the wire number can then be used to sort the list back into the original order (if you're using a spreadsheet). Sorting by connector can be useful as you will then notice right away if any pin is not used or is duplicated, which is a common problem.

The last column provides a space for ticking-off the wires as they are laid in to the loom. Where several people are attacking a single large cabinet, a single stroke '/' can be used to signify that the wire in question is being attended to, followed by another stroke 'X' to say that the wire is in place. Additional columns for wire lengths might be useful.

A table like Table 3.2 should be accompanied by notes explaining what J1, J2, etc. are, along with mechanical layouts which might clarify their positions inside the enclosure and which will provide clues as to the lengths of wires.

It can be awkward visualising the layout of a list like this. A good idea might be to produce a single example diagram which shows just one set of connections for one control as an aid to understanding. I've left that as an exercise for the reader.

Systems mounted in racking enclosures, where much of the wiring is in the backplane p.c.b., may not need an interwiring diagram or may only need a simple one covering power supply arrangements. Simulators and similar systems may need large amounts of wiring.

If you generate any large wiring looms, feed this information back into the mechanical layout diagrams to check that there is really room for them and to see whether there are convenient attachment points where the looms can be hung up. The actual route of wiring should be drawn into the mechanical layout.

There is an important variant on the interwiring list, the cable list. Where you have a number of cabinets joined by cabling then a cable list makes it easy to specify the cores used. It can be similar in form to an interwiring list.

Circuit diagrams

The circuit diagram is the one document which everyone feels is the 'real electronics'. Developing the actual circuit is seen as the nitty-gritty design process.

The circuit diagram is the other entry point to the resolution phase of a project. The only drawback to tackling circuits before wiring is that you need to think about wiring as part of the circuit rather than as a separate issue.

The advantage of tackling circuits first is that issues about the precise nature of switches and other electro-mechanical furniture may resolve themselves sooner, meaning that there are no blank spots on the interwiring documents. Also, heat sinking requirements may have been calculated during circuit design, meaning that these possibly bulky items will not be unknowns when the panel diagrams are being drawn up.

Without going into actual circuit design for the moment, there are still points which can be made about the diagram itself. Completeness is the most important criterion. All components must have a value or type quoted. Where a particular technology, tolerance or grade is required but is not implied by the type number or is not the 'usual', it must be mentioned explicitly.

For capacitors, I tend to use the shorthand PE, PS, CE, SM and so forth for polyester, polystyrene, ceramic and so on. For low voltage circuits, capacitor voltage ratings are not relevant, except in the case of electrolytics whose voltage rating is probably important.

The development of circuit diagrams ought to be accompanied by a program of testing and modification which results in real working fragments of the system. There will be natural sets of blocks which can be tested together, along with natural breakpoints where test signals can be conveniently injected.

Quote pin functions to tell you what the circuit is supposed to be doing while you're testing and breadboarding (these tend to go inside the symbol). Quote pin numbers to make p.c.b. design easier (these will go outside the symbol or box and alongside the connection). Where there are several gates or devices in the same package, defer pin numbering until you see what the p.c.b. tracking is like and then choose the most convenient combination of gates for the layout, remembering to transfer the information back from the p.c.b. to the circuit diagram.

Figure 3.7 shows a typical circuit diagram.

Placement charts

At some point you will need to decide what is going to be on each board. Until you do this, you will not be in a position to finalise or 'firm-up' on much of the mechanics. A lot of smaller projects will fit quite happily on a single board; sometimes it will be obvious when this is the case. If this is so, then fine. Where this is not the case, a physical re-partitioning of the logical partitions in the block diagram is required, resulting in several larger partitions, each of which represents the contents of one board.

As we gradually derive a set of circuit diagrams, we can begin to develop a sense of how physically large individual parts of a circuit will be and how much space or area will be taken up on the board. Exact areas will need to be confirmed by the placement charts and the p.c.b. designs themselves, of course. If we have our wits about us, though, then any blocks which translate to significantly large numbers of real components will set alarm bells ringing and will cause us to go back to the physical layout. Once there, we can answer questions such as 'have I got room to stack several boards if it should turn out that I need to?'

Apart from their usefulness in predicting whether large board-mounted components interfere with panel components and such, placement charts tend to be very much 'after the fact' documents. Once a verified circuit diagram exists, a first approximation spacing requirement could be worked out. This will not be exact but you can gain some insight on the basis of a fairly simple calculation. You might even be able to decide whether several boards are going to be needed or if one will suffice.

To make an approximation to the required space, look at the circuit diagrams and count the number of resistors. Then count the number of *pairs* of pins on 0.3" pitch i.c.s and the number of *pairs* of pins on 0.6" pitch i.c.s. If there are not too many capacitors, they can count as resistors for this purpose. The rare 0.4" pitch i.c.s can probably count as 0.3" pitch i.c.s.

Now modern resistors use a spacing of 0.4" between pins; 0.6" allows for the size of the pads and for space in a lengthwise direction. Resistors probably need lateral spacing by 0.15" so the board area occupied by each resistor is 0.6" × 0.15" = 0.09"^{2}. You might get away with 0.2" between i.c.s so each pair of pins uses 0.5" × 0.1" = 0.06"^{2} for 0.3" pitch i.c.s, or 0.8" × 0.1" = 0.08"^{2} per pair for 0.6" pitch i.c.s.

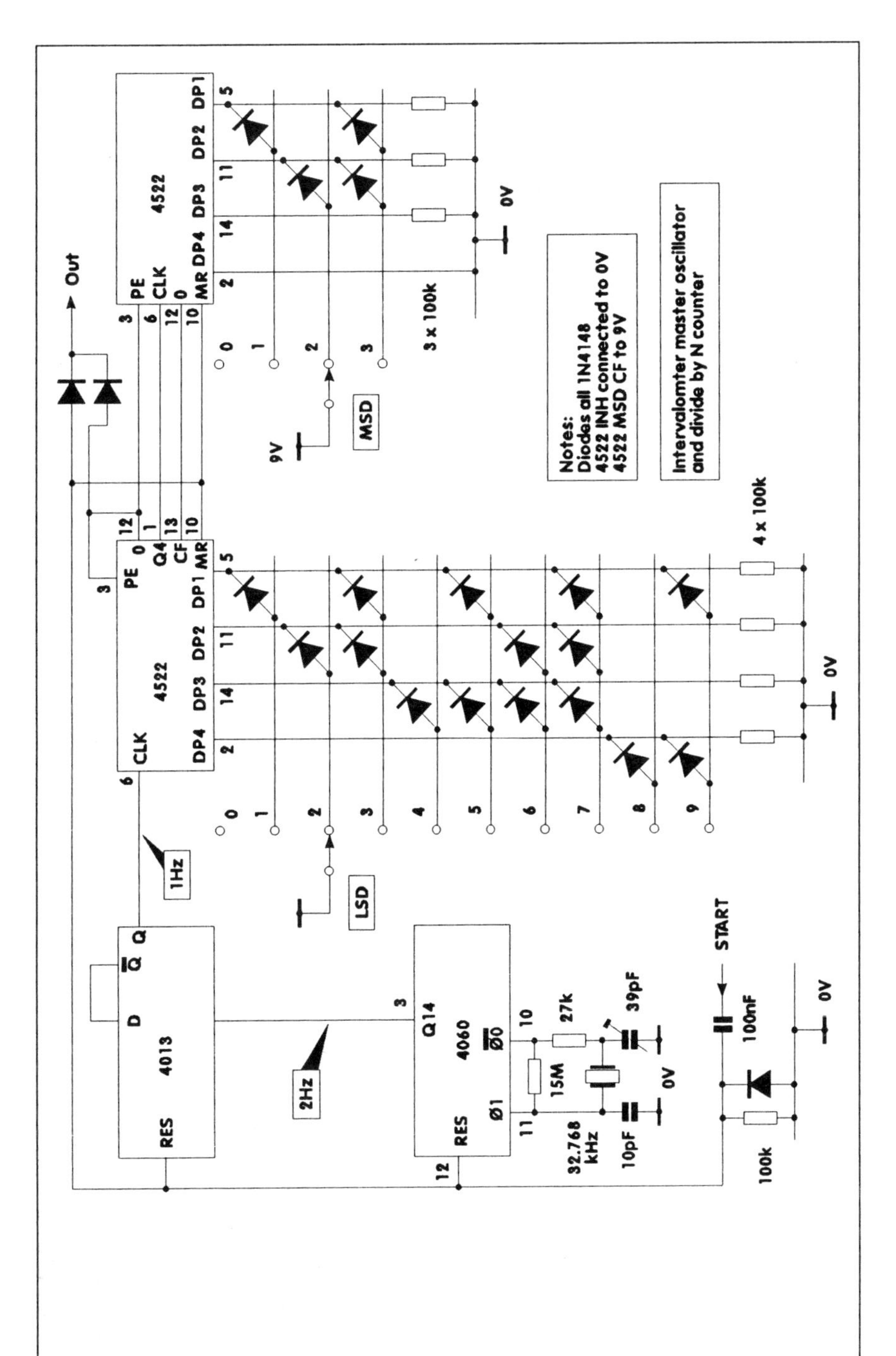

Figure 3.7: A circuit diagram

Multiply the area for one resistor by the number of resistors and do the same for the pairs of i.c. pins. Add all this together. Also add the area due to any large single components. Don't forget to allow space for connectors, or the equivalent in pads if soldering direct to the board. To use some simple algebra:

$$A = (0.09 \times r) + (0.06 \times b) + (0.08 \times c) + a$$

where r is the number of resistors, b is the number of pairs of legs on 0.3" pitch i.c.s and c is the number of pairs of pins on a 0.6" pitch. a is the area of any other one-off items which the board needs to accommodate.

I use tenths of an inch rather than millimetres since this is what we've inherited as a pitch between i.c. pins. Of course you can re-jig the calculation for the sizings and spacings which you prefer, or you can experiment to see whether different spacings materially affect the number of boards.

Multiply the length and breadth of your prospective board to give its area and compare this with the result from above. If the required area is less than 75% of the available board area then there's a good chance of getting away with one board, unless you've forgotten something. If the required area approaches 100% then you will need two boards unless you can squeeze things up. 150% implies two nicely spaced p.c.b.s, and so on.

As I say, this is an approximation and needs to be confirmed by actual p.c.b. design. Remember that the overheads are proportionally larger for smaller boards. The smaller the board is, the less chance there will be of juggling the layout and squeezing things up.

If you feel the need to use a computer for this, maybe you're going to do a lot of juggling with the p.c.b.s, then you could use a spreadsheet to do the calculation or write a little something in BASIC if you like.

This calculation outlined above is likely to be more appropriate for analogue electronics where tracks tend to run rather conveniently under passive components. Those areas of board devoted to digital circuits tend to need a certain amount of space for tracks alone, unless a multi-layer board is being considered.

Of course if you're going to use surface mount devices (SMDs) you'll pack the components into a far smaller space. The same principles apply, however, and you could do a similar calculation based on the sizes of SMDs.

Much of the information on a placement chart can be obtained from a p.c.b. design, once this is completed. The placement chart finds its true role in life as part of the build manual or trouble-shooting guide. Incidentally, there may be other reasons than those of pure space for using multiple p.c.b.s; we'll give those an airing under the heading of partitioning.

Figure 3.8 shows a typical component placement chart. The numbers round the edge are usually used if the board is too dense to fit on the component designations; in that case, the chart is accompanied by a table showing the grid locations of each component. In the case of Figure 3.8, for example, both R3 and R4 would be placed at about E3.

Often, the placement chart is generated as a by-product of the actual p.c.b. design process. In that case, it is often used to overprint the finished board. This is a boon to the production side of the operation, where a misplaced diode or resistor can result in an expensive re-work.

A word on component designators. These need not be in any order; the main thing is that they be unique, at least across any given board type. It is not strictly necessary to have no missing designators (e.g. an R1 and an R3 but no R2); after all, we may at some time want to delete a component. You can either apply component designators to the placement chart and copy them to the circuit diagrams or vice versa.

Component designators are most usefully referred to in the guides and manuals written during the resolution phase. Occasionally, they form part of a parts list.

You may need a rather different kind of placement chart if you're using wire wrap or stripboard as a construction method. The same means of calculating probable areas can be used, although component packing density is likely to be higher on a wire wrap board than on a p.c.b., especially for digital circuits with few passive components.

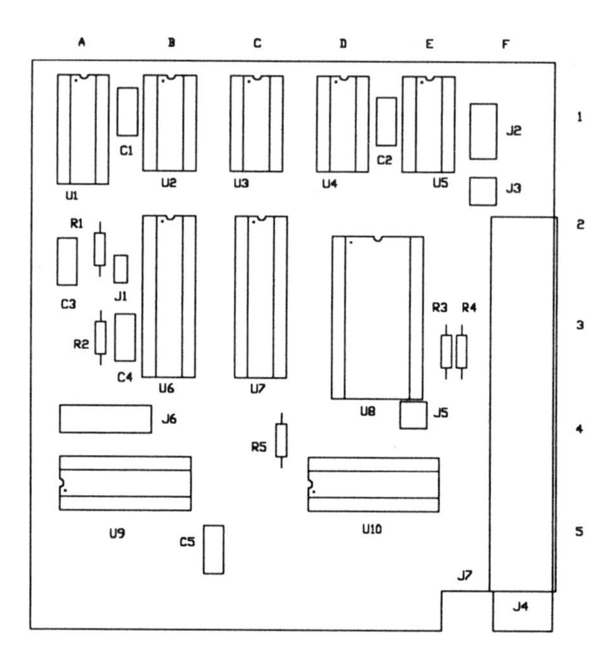

Figure 3.8: Component placement chart

P.c.b. foils

Little needs to be said about p.c.b. artwork itself at this stage. There is very little to it apart from the tracks and pads themselves. The black arts of p.c.b. design are given their own section. Although p.c.b. artwork is very important (if you're using p.c.b. construction) its sole use is in making the actual p.c.b. It is not much used anywhere else. Speaking generally, completeness is the order of the day for p.c.b. artwork, as with all documents at the resolution phase.

Parts lists

The parts list, despite its relative simplicity, is one of the most difficult things to keep properly up to date. It seems to insinuate itself into everything. Every decision we make reflects to some extent on the parts list. Maintaining a parts list can be tedious; although ripe for automation, there is very little chance of automating it *thoroughly* without really sophisticated and expensive computer software. The best CAD systems spit out a 'bill of materials' at the end of the design process which can be used as the basis for a comprehensive parts list.

Unless you have access to these rather specialized systems, the best plan is to take another look at the list every time you make substantial changes to any other part of the system. Take notes about potential changes and spend ten minutes at the end of the day making the alterations to the parts list. Split the list so that the parts for individual boards or subsystems are in individual columns or on individual sheets so that you can attribute the parts to their uses.

We could just have an 'after the fact' or as-built parts list, but this doesn't help us to order the parts we need to do the experimentation and development.

Recently I've taken to having two parts lists. The first has all the components on it that I need right away for testing the actual circuitry. This minimalist list is an educated guess derived from a scribbled circuit diagram; there will be few if any mechanical components in it. It is a safety measure in case I need to do any re-working which might mean that some parts were ordered in error. My bits box might also contribute some components for this first tranche.

If I've forgotten anything on the first list I can order it the second time around; it is less likely that there will be small orders or afterthoughts which simply swell the profits of the parcels carrier. It isn't a first approximation parts list in the true sense since it does not pretend to be complete.

The second parts list is an extension of the first list and tries to be complete if it can. The trick is (for one-offs at any rate) to keep back the components used in testing and then order any excess required from the supplier. Eventually, you will end up with a parts list which can be used by the production effort to order what is needed to keep production swinging along. Such a parts list needs to be scrupulously accurate, especially where large numbers of units are being thought of.

Once you're sure the prototype is good, use it to check the parts list over.

There is a case for ordering some *examples* of expensive mechanical parts early in the cycle, particularly if these are unfamiliar to you. There is nothing quite like actually handling an unfamiliar mechanical part to give you an appreciation of its size and other qualities. You can measure it and 'offer it up' (as the joiners and woodworkers say) to its intended position and see how it really fits. Sound psychology.

A parts list which is intended for use in a manual rather than as a basis for ordering parts may not group identical parts together but may instead split identical parts into individual component designations. It used to be the fashion to give perfectly ordinary components one's own part number and provide a list of relatively expensive spares at the end of the service manual.

Table 3.3 shows an example of a parts list, shown as part of a spreadsheet.

The two columns marked 'HLS' and 'TDM' are supposed to represent two parts of the same project, two different boards perhaps. Doing things this way helps to keep things together without mixing up the parts for different subsystems.

Note the need to order minimum quantities of some items. A lot of smaller components are sold in packs containing perhaps five or ten items or 'standard supply multiples' (SSMs). Just to add to the confusion, SSMs are individually priced whereas 'packs' in the true sense of the word tend to be priced per pack. So look out.

To get an idea of the true cost, the individual price must be obtained by division by the pack quantity and then multiplied by the quantity to be used.

Description	Suplr	Part No.	HLS	TDM	Total	Pack	Order	Each	Cost
BC107	RS	300-095		6	6	10	10	0.173	1.73
Res 22k	Fnl	MR-22k	2	11	13	10	20	0.020	0.40
Res 33k	GEC	002-33k	2	1	3	10	10	0.020	0.20
Board1			1		1				0.00
Board2				1	1				0.00
CMOS 4013	STC	HFD4013	2	4	6	1	1	0.220	0.22
CMOS 4060	RS	300-096	1		1	1	1	0.450	0.45
Cap 22n PE	Map	CAPE22	3	2	5	5	5	0.080	0.40
Bolt M3x6	Map	09-0045	4	4	8	10	10	0.015	0.15
Totals									**3.55**

Table 3.3: A parts list

Ordering must take account of how many items there are in a pack or in an SSM.

Table 3.3 is a simple example in which a great deal has to be done manually. I have designed spreadsheets myself to automatically use price breaks and existing stock levels, but I don't propose to go into such detail here.

The cost of an order will not be the same as the cost of the components in the finished product. Table 3.3 shows the cost of an order, not the finished project.

More sophisticated spreadsheets might even have columns for alternative suppliers. The parts list is a real candidate for computerisation by spreadsheet, even if we cannot automate it completely.

Power supply requirements

Where power requirements are fairly heavy, or where a number of items of equipment are to be powered, a list of power requirements will help you to choose a suitable supply or supplies. Tot up the currents drawn by each piece of equipment under full load at each supply voltage. (This is always providing that the supplier can tell you what the power supply requirements of his products are; oddly, I've known manufacturers who couldn't even tell me what voltages their equipment operated on, let alone the current requirements.)

Don't cut corners with supplies rated at lower currents unless you're *certain* that not all equipment will be powered up at once. By 'certain' we mean there is some switch or hardware interlock which is *always* present and which makes it *impossible* to switch everything on at once. This business of a 'diversity factor' or current requirement derating is all right in lighting and domestic mains power where the biggest disaster is a blown fuse and a spoiled soufflé, but it needs to be 100% for the things we do.

Diaries

A diary might be a useful informal addition to the documentation. We've already talked about keeping notes to transfer between different documents; these are, in a sense, diaries too.

There's no need to do Sam Pepys impressions, just jot down brief details of your progress and any minor technical items, phone numbers, appointments and so on as they crop up. Invariably, a lot of scraps of information will come your way which will not fit easily into the more formal documentation. A diary, scrap-book or notebook is a good place to keep these useful snippets.

And finally . . .

Use your imagination; keep things up to date and keep them complete. It may seem like a lot of work to start with but, as I said before, you'll benefit in the long run by having everything to hand when it's needed.

Everything is built and tested at last . . . great! Now go and look at the documentation again and make sure that it really matches the prototype you've built!

4: Tools for Design

In a very real sense, documentation is a tool itself, a kind of DIY tool which mirrors and records your design activities. In this section, however, we're looking at physical 'tools proper' in their many forms.

Small hand tools and other items specifically intended for construction activities are covered in Chapter 10.

Brains and other biological bits

The most useful and versatile tool you can possess, bar none, is your own brain. It is the one thing which makes all else possible and it is far more capable than most of us give it credit for. This is not a book about biology, or about psychology as such, but the poor brain comes in for so much abuse that it's time to set the record straight.

Reading this book is probably a learning experience for you. You probably want to retain the information in it and have access to it on a moment by moment basis. My aim in this short section is to show you a few tips which will allow you to retain information more easily, use the working environment in a productive manner and stay motivated when faced with adverse conditions. Sometimes successful engineering depends more on these factors than anything else.

The facts: your brain is a parallel network computing system consisting of about ten thousand million interlinked computing elements (neurons), all active simultaneously and running at some millions of millions of connections per second (which is how these things are measured for a network computer). It is worth far more than any silicon machine with perhaps four million storage elements and a single processor cranking over at twenty-five million revs, or sixty-odd million if you're lucky.

Now I'm a bit of a computer buff and I use the silicon version all the time, even at this instant I'm typing this sentence right in there, but even so I'd rather keep my wits than have a shiny new PC, if I had to make the choice.

Where silicon is the workhorse of automation, the brain is the tool of the imagination. Much of what a human being does doesn't depend on winding a handle to get a routine answer out. That's the silicon way. Instead, human effort and success depends on the use of advantageous features of the environment. Silicon itself is one of those advantageous features.

The role of silicon is to crystallise the results of your imagining, at most, to assist with your powers of visualization. It does not replace your imaginative powers. Too many people expect to wind a handle and that some solution will pop out of a machine with no mental effort on their part.

I'll repeat myself: design is a human activity whose end product is itself destined for use by human beings. Before any handle is turned, any printout is printed or any design is designed, human imagination and motivation must be switched on.

The question is, how to keep the old brainbox running at peak performance. The care and feeding of brains is an important subject. Despite its amazing power, there are one or two disadvantages to having a brain. For one thing, we have to keep it motivated. We need to feed it properly, avoid bashing it against things and provide it with interesting and informative inputs. It needs to rest every now and again (even while we're awake). Although there are a few remarkable exceptions, it tends to work better when supported by a fit body.

Some tips: it may concern you to have discovered that you can't hold more than a few things, seven or eight at the most, in your mind all at once. Don't worry about that; even the cleverest people can't. They may appear to do so, but they're actually structuring their memories into linked or hierarchical groups which remove the need to think about so many things at once. If you sincerely need to remember dozens of details like that, then practise structuring your thoughts using a spray diagram or similar.

Otherwise, do what the rest of us mortals do and keep tidy notes as an *aide memoire*, then use them as frequently as necessary. Don't make lengthy text notes and then attempt to memorize them; you might just as well have tried to memorize a whole book. An example spray diagram is shown in Figure 4.1.

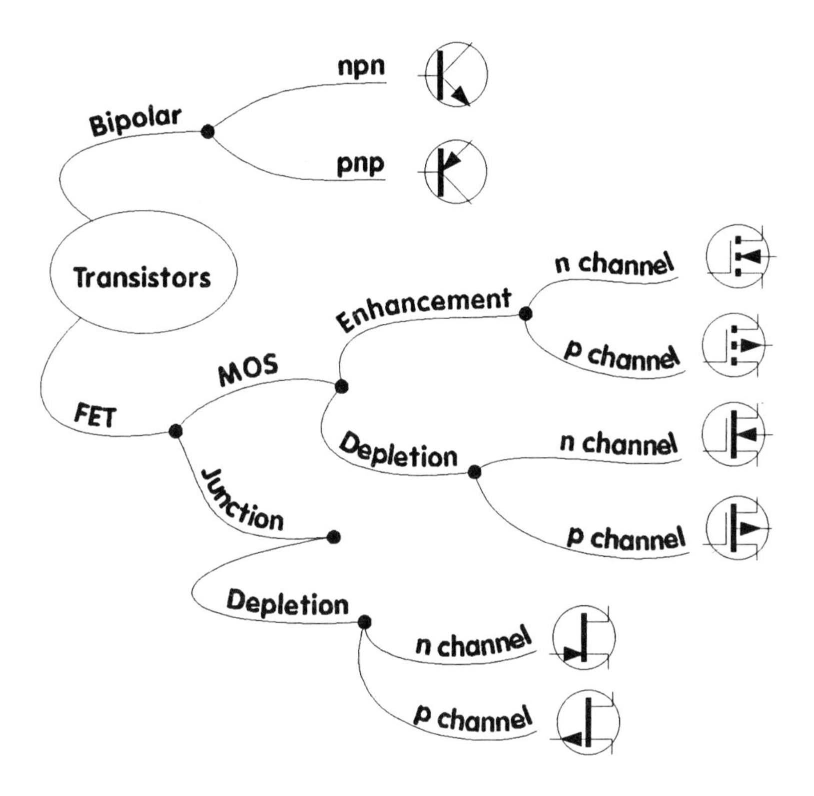

Figure 4.1: An example spray diagram

The attention span of most people is about forty minutes to an hour. Take a break at about that time interval or at least switch your attention to some other aspect of what you're doing. A change is as good as a rest. Everybody's different; you may last a little longer than this or perhaps for not as long.

It is interesting to contrast the usage of the brain between someone who is proficient in the task at hand and someone who isn't. We use certain parts of our brains for skilled work and other parts for stumbling along the path of learning. Rather than there being a part of the brain devoted to 'designing electronic gadgets' and a part for 'cycling along forest tracks' there's a part of the brain for 'I'm good at this' and 'I'm not so good at this other'.

If you're learning something, do it in private where possible until you've got it taped and can put on a 'performance'. Practice of the productive kind gradually transfers the thinking to the 'fluent' part of the brain where it happens automatically. This is especially true of manual or motor skills but it applies to a large extent to mental skills too.

If you're studying, get the cues right. Don't expect to learn much in a tense or stressful situation, except that learning is a tense and stressful business! This is 'inappropriate learning', associating two ideas which are not necessarily associated.

To stretch this point a little further for students. Ideally I suppose we would study in a room similar to the one we're taking exams in. This isn't always possible. A good substitute is to visualise yourself happily working away in there. Visit the examination hall and get a good look at it, if possible. When you finally arrive in the examination room, you won't be a bag of nerves - you'll know what to expect, you'll have been there before, albeit in the imagination.

Inert knowledge is knowledge we can't access because the cues are incorrect. We learn something in school, say, but it doesn't relate to what we're doing now. There are plenty of examples in psychology of inert or suppressed knowledge which only comes to light in the peculiar setting in which it was first learned. Avoid accumulating inert knowledge by visualizing using your new skills in imaginative situations. In this way the information becomes integrated and it will be at your fingertips (or is it neuron-tips?) when you need it.

That reminds me of the tale of the laboratory technician. At five in the afternoon, off came the lab coat and out of the door she went. Asked what she'd done that day, she would not have been able to give you an answer. If you'd had a spare lab coat handy, you could have had an answer, because as soon as she popped the coat back on it all came back to her. This is a rather extreme example but it does illustrate the value of cues.

The key to much of motivation also relates to use of the imagination. Imagine how the user uses the product. Visualize yourself having to access and repair the system. Imagine how the gadget will behave when it's finished. Keep yourself moving towards the target by keeping these things in mind.

Productive use of the imagination can be learned. It just takes practice. As you practise, your ability to focus sharply on particular aspects of a design will be enhanced. Your ability to find innovative solutions will be enhanced too. Remember to jot down the results of your imaginings on the documents and forms which I described earlier.

Imagination does not mean living in a fantasy world, it means bringing the results of imagining down to the real world. Take the time to sit still and think for a bit. Don't be afraid to change. In the long run, it's worth it.

A toolkit for the mind

On a subject related more directly to electronics, I developed an idea which I call the 'toolkit for the mind'. Sometimes we don't have access to an instrument that will do the job we want. Except in the stranger science fiction novels, there isn't a device that we can apply to a circuit diagram (a piece of paper, in other words) to measure an actual voltage or current. Perhaps we could apply the latest SPICE to a model of the circuit, but there's another approach we can use first: the toolkit for the mind.

This toolkit consists of a set of imaginary instruments and tools which you can apply to a system on paper. Basically, while poring over a diagram of some sort, imagine what the imaginary DMM would read if put across *here*, say or *here*; imagine what the imaginary scope would look like if applied to *this* point. It works for mechanics too; imagine how the imaginary screwdriver can be insinuated into *that* corner.

It's a useful trick and apart from its other advantages can give you an insight which is difficult to gain from measurement of a lone sample. The instruments are of course perfect; they are capable of making any measurement and don't influence the circuit under test at all!

I find that using this 'toolkit' helps me to understand how things work in practice. The only down side is that you'll need a certain amount of experience with real instruments and tools before it will work well for you. But try it anyway; it's useful even to try.

As I've said before, this is not a book on psychology, so I've been brief. But I'm amazed by how much of learning and design is given over to chance or brute force repetition, so I've taken the opportunity to lecture you a little on the subject.

So far as bench work goes, eyes and hands need looking after and using properly. Eyes in particular seem to come in for some stick. I've a friend who claims that breadboards drive him nuts; the multiplicity of wires confounds his eye. I have no problem with breadboards but must confess that wire wrapping drives me cross-eyed. Proper lighting helps, of course; lighting is discussed in a later section.

If you're stuck at the bench, take the opportunity to jump up now and again to go for a stroll and a chat. Slowly stretch the neck, hands and limbs, especially if you've been doing a repetitive job or standing or sitting still. A change of posture helps. If I spend too long in one position it creases me up; I suspect that it's the same for most people. A blast down at the pool or the gym is a good fix.

Computers: help or hindrance?

A computer can help you or hinder you. A computer will not organize you, but it can help you organize yourself. Computing can be a frustrating business, and the results of your endeavours at the keyboard, especially if they involve a numerical model, need to be interpreted properly. There are numerous horror stories about people taking the results of some modelling work as gospel and the awful consequences of that. So don't hide behind the computer, use it properly.

A computer will help you to do the routine things which are perhaps tedious and which involve making painstaking calculations which are prone to mistakes. It will help you record things and keep them in order. Coupled with a decent laser printer, it will conveniently produce presentation quality artwork if your drawing skills are not fantastic. It will enable you to send and receive documents and drawings in a machine readable form, provided the recipient has software compatible with your own.

A computer will hinder you when it needs to be set up differently for different purposes or when it has broken down and you have lost the use of it or (worse still) lost your data. It will hinder you when you have to fiddle with it to get it to do what you want, or where the results of a computation need to be interpreted strongly to get at the true meaning.

Much software requires that you learn its own idiosyncratic way of doing things. That can be a pain. On the other hand, learning about three different spreadsheets or five word processors can give you a very broad and useful understanding of the whole field, if you have the time for it, enabling you to sit down in front of any version of a sensible software product and make it go.

The computer is just a tool and is only as useful as the time it saves. Mine saves me a lot of time because I've got it fixed up to my needs, although it did take a while to get it to that state.

There is a current theory that desktop computing power has not increased productivity one jot. There are a couple of good reasons for this. First, people may spend too long fiddling getting things to work the way they want rather than bashing on with what they should be doing. Second, it's tempting to justify the computer's existence by doing trivial things on it which might have been better done by hand.

A computer is a great tool for automating the organization and storage of the details of a project. Note, however, the emphasis on *automation*. To automate properly, you need to be set up properly and things need to fall to hand right when you need them. If it takes you longer to get set up or customize the system than it saves you in time and effort, then automation is not worth it; go back to pencil and paper. Automation need not be applied to all aspects of a project anyway.

My comments below apply especially to IBM compatible PC/AT-type computers but also generally to other types as well.

The care and feeding of computers

A few words on the care of computers. First of all, and this might sound silly, computers do not like being moved. Throwing bits of computer in and out of car boots and back seats or shuffling computers around offices is not calculated to keep them healthy. Things get scratched and buffeted; they work loose and get lost. Connectors work and slacken a little. Move your computer around as little as possible. If you are mobile, get a proper laptop or notebook. If you're forever unplugging and plugging in printers, get a printer switch or another printer.

Keep food and drinks away from the machine! Also, keep media (disks, tapes) away from sources of heat, moisture and strong magnetic fields. Incidentally, airport X-ray machines don't harm disks - not in my experience, anyway.

Once you've got the computer to a workable state, make as few changes as possible. Every enhancement, new board or software installation is a potential booby trap. If you must make changes, make them one at a time and check that everything continues to function as before. If you've been observant and you know how your computer normally behaves, then you're one-up at this point.

Exit software before switching off. Idly flicking the power off while an application is running can not only corrupt data (the disk write buffers may not be flushed until the software exits) but may also leave a lot of temporary files hanging around on disk, taking up space.

Data costs more than the computer ever did. After all, you may have spent months typing, drawing, saving, collating and so on. If you don't want to spend a few more months doing it all over again, then back up regularly.

Keep data separate from programs, otherwise you'll have to untangle the data files from the application files when you want to back them up, or face the prospect of backing up the whole application along with the data! Ugh! With modern applications this untangling is difficult; not only are there more separate files devoted to an application than there were a few years ago, but many of these files use names or extensions which are similar to the data files which they create.

I have literally lost count of the number of times I've run into situations where data and programs were kept together, even to the extent of having a complete word processor mixed with letters and invoices all in the hard disk's root directory. An arrangement like that is almost impossible to unravel.

There was some excuse for this with primitive and archaic DOS programs of five or more years ago, which imposed restrictions on where data could be kept. Any modern program worth its salt will make it easy for you save data where you want, so take advantage of that.

My own system has a directory, called DATA, which contains sub-directories for everything else. This is a fairly sensible arrangement as I just need to back up the DATA directory to be safe. The other major top-level directories are DOS, WINDOWS, DOSAPPS and WINAPPS. Everything neatly kept on its own patch.

I used to keep all my data on floppies; eventually, that had to go by the board. Some of the files generated by modern applications are huge and will literally not fit on a floppy. Floppies are a lot slower than a hard disk too. The hard disk is a convenient place to keep data as it is all together (physically at least) and, in principle, instantly available.

But you pay a price for the convenience and speed of keeping data on the hard drive, that of vulnerability to loss. Actual failure of the hard drive is less of a problem than operator error, often one person compounding a situation by making incorrect assumptions about the storage location of someone else's work. On those (rare) occasions when I've lost something it's been my own silly fault. I couldn't blame the hardware or the software.

Backing up to floppies eventually became a pain as my DATA directory was getting bigger all the time, so I eventually got a tape drive, which has been marvellous. Late on in the afternoon it just hots up and does a back-up all by itself. I back up the whole DATA directory, plus a few other things, on to two data cartridges, alternating between them.

Backing up is vastly more convenient using a tape back-up system. For one thing, you don't need to stand over the computer armed with a batch of floppies, ready to feed them into the drive, although I still make a plain copy of the files of particular projects to a floppy or two now and again just for added peace of mind.

You might also want to back up customizations along with your actual data. Modern software tends to keep customization data in a separate file; in particular, look out for files with .CFG, .INI, and .CNF extensions. If you've spent some hours getting your favourite word processor installed, tuning it and turning various features on and off, then saving customizations might be an option you would like to take up.

Again, modern tape software allows you to include in your automated back-up just those files and no more. Alternatively, you might want to keep a paper record of what you've changed so that if you need to re-install you can get things running more quickly.

You could use the DOS xcopy utility to copy any customization files that have changed into, say, a CUSTOM sub-directory of DATA. When you back up the DATA directory, all sub-directories, CUSTOM amongst them, will be backed up (remember to use the /s switch with your back-up command or whatever). If re-installation is ever needed, you may be able to fool the application into using your previously backed-up version of the configuration file. This latter method, unfortunately, requires that you be very familiar with the inner workings of your software; there is no space here to give details of all the possible ways of doing that.

All this makes computers sound like a minefield, doesn't it? They can be if they're mishandled. In fact I obey my own rules and I find that my own computer is available 99% of the time. The other 1% is usually down to the local electricity supply company. If you're plagued with power cuts, get a static inverter unit. I've seen respectable makes on sale for less than £100 recently.

Useful software

60% or more of what the world does on a computer is word processing. Obviously you can write letters and reports and manuals on a word processor. But don't be shy of making up lists of jobs or even parts lists on a word processor (although a spreadsheet is better for the latter).

In describing how software works, I've tried not to give detailed instructions for a particular product but have instead given general comments which should be applicable to all that class of software. Please, read your own user guide for more details.

Spreadsheets

I use spreadsheets a lot. They are the most versatile piece of software there is. Traditionally, they have been the province of the accountant and his ilk, but modern spreadsheets have all the things which an engineer holds dear, even to the extent of supporting trigonometry and matrix operations.

From the point of view of the engineer, the activities which you can automate using a spreadsheet fall broadly into two categories: technical engineering and engineering administration. Any assistance with the numerical end of the design process (analytical design) falls into the first, whereas the second category principally covers that of maintaining parts lists.

First, a brief look at how spreadsheets work. What you'll see to start with, immediately after you've loaded up the spreadsheet application, are rows and columns of empty cells. The rows are probably numbered and the columns may have a number or a letter to identify them. Cell *A1* (or *R1C1*, row-one-column-one) is always the one at the extreme top left of the sheet.

You can type one of three things into a spreadsheet cell: some text, a number, or a formula (an 'expression', if you're a mathematician). Although most spreadsheets actually have functions for chopping text up or joining it, for our purposes text is merely used as a title or to name something and will not need to be manipulated in any way after it has been entered into the cell. Numbers are numbers and anything which looks like a number will be classified as such and stored by the spreadsheet as a number.

Formulae are a little more complex. Check your spreadsheet documentation; you will probably need to use an = sign ahead of the formula if it is to be interpreted as a formula and not just as text. The numbers used in formulae can be constants or they can be references to the results of previous calculations contained in other cells. When you have entered a formula into a sheet, you do not see the text of the formula appear in the cell. Instead, you see the result of the calculation which the formula has done, unless you have turned that facility off.

For example, if the cell *A1* contained the number 0 and the cell *A2* contained the formula `=COS(A1)+0.5`, then what you would actually see in cell *A2* would be the number 1.5. Usually, a formula is re-calculated as soon as you finish entering anything which affects it. Expect to see a sheet change in various places whenever you enter data. The underlying formula is usually shown on a special status or editing line above or below the sheet itself.

Once you have developed a feel for the way spreadsheets operate, you'll find it's very easy to quickly knock up small sets of formulae.

One little gadget which I found very useful recently was a calculation for determining the resistances needed to give me a trimmable potential divider for setting voltage references accurately. It would not have been worth it for a single circuit, but as I was faced with the prospect of fiddling with a dozen of them I decided it was a good idea. The circuit is essentially a potentiometer between two padding resistors. Figure 4.2 shows such a circuit, which is an important sub-circuit element in many larger circuits (I will be mentioning this kind of circuit at length in the section on designing for adjustment).

Table 4.1 shows a spreadsheet which models the potential divider circuit of Figure 4.2. I enter into the sheet the upper and lower voltages to which the ends of the potential divider are connected (V_U and V_L in the Figure). I also enter the current I want to flow in the chain, and the voltages on the ends of the potentiometer (V_{VU} and V_{VL}). (I could have designed a spreadsheet to enter the desired mid-point setting of the potentiometer and a desired percentage available adjustment if I'd wanted.)

The sheet calculates four resistances: a total resistance, one for the potentiometer and one for each of the fixed resistors. These will usually be peculiar or awkward values, so there is a second section of the sheet which works the other way about. Into this, I enter plausible standard resistor values, guided by the results from the first calculation, and it shows me how good an approximation I have to my ideal divider chain. I can then play around

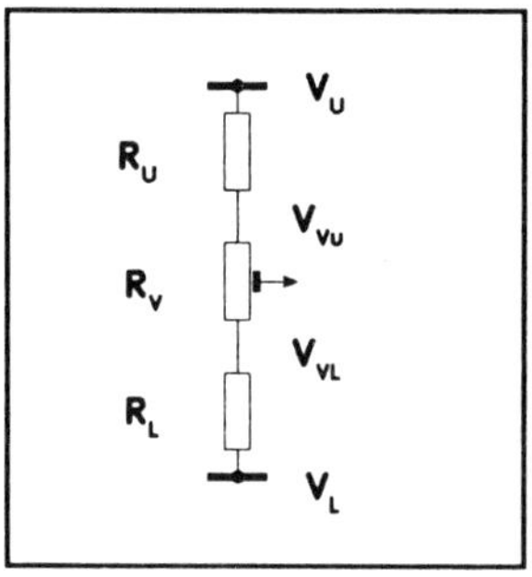

Quantity	Forward	Reverse
+ve end of chain	12	12
-v end of chain	0	0
+ve end of pot	8	7.9655172
-ve end of pot	7.5	7.4482759
Chain current	1.00E-04	1.03E-04
Total resistance	120000	116000
Upper resistance	40000	39000
Lower resistance	75000	72000
Pot. resistance	5000	5000

Figure 4.2 & Table 4.1: Potential divider circuit and spreadsheet model

with standard resistor values until I get something I can live with. Using a calculator takes *ages*.

This is a simple sheet and I have not bothered to protect against entering negative resistances and other sillies since it would not be worth the time involved. Such mistakes ought to be obvious on a small sheet. The sheet took me about half an hour to perfect and it saved me several half hours as a result.

I've used similar techniques in the past to design filters and to investigate the results of fabricating transistors with different doping levels and base region widths, amongst other things. If you can turn it into numbers then a spreadsheet will almost certainly handle it.

One trick to use, certainly on the smaller technical spreadsheets which tend to look a little scrappy and disorganized, is to distinguish firmly between areas of the sheet into which you may freely type (i.e. inputs), areas where results are to be displayed and areas which are purely titles.

I do this by marking the cells with different background colours (even on a monochrome system you could try shading). You then don't need to start using cell protection to prevent yourself from entering data into inappropriate cells. Blue is good for permanent titles; green (for 'go') is a good choice for input areas and yellow is fine for drawing the attention to cells in which results are to be displayed. Some background colours may make the cell contents hard to read.

Discrete time step analysis

You can do simple finite element or discrete time step analysis using a spreadsheet too. While these are not the last word in ease of use or speed, you can get some good insights into circuit behaviour. An excellent by-product is the graphing facility which most spreadsheets have. Analogue circuits are the best subjects for this treatment; digital circuits, with the exception of digital signal processing (DSP), which is a good candidate for spreadsheet work, are best treated in other ways.

To do time step analysis, make the first column of your sheet a timing column. Reserve the top row for a title, then enter integer numbers starting at zero, down the column. Some sheets will have automated methods of doing this; Excel's `Data..Series` is a case in point. The numbers could, if you wish, represent actual microseconds, milliseconds, or whatever. There are no fast rules about how many rows you need, but at least twenty is a good idea. You will usually need more.

The stimulus column comes next. It will contain the value of an input signal at the given time instant. Enter an appropriate formula in the second row. As an example, `=SIN(A2/10)` is probably the simplest. The formula must refer to the timing column in some way; here, we are just using a simple sine wave which starts at zero and repeats about every 31 rows. Copy the formula down alongside all the timing column, in whatever way your sheet allows.

Arrange for each interesting point or value in the circuit to have its own column. In the top row, type in its name. In the second row, enter numbers which represent the starting conditions of the circuit, where these need to be used by subsequent calculations. The initial voltage across a capacitor is an example.

Derive a simple algebraic relationship between each value of interest and those in preceding columns; enter these in the third row. Make up one row and then copy it down the sheet. If you've done everything right, the calculations should show some significant behaviours. Make the time steps sufficiently short, otherwise you'll run into that well-known sampling bugbear, aliasing.

Play with the sheet by altering the values in the formulae. It would be a big problem to type all these in, but the ability to copy whole rows removes the need for lots of typing. We've rather glossed over the use of the spreadsheet in this situation, but I hope that you have grasped the general idea. Remember that you'll need to test any circuit whose values you've derived in this way, in the usual way for any circuit. Don't take computer figures as gospel.

I dare say you could do frequency domain analysis in a similar way if your math is up to scratch. In fact there are plenty of frequency domain analysis packages on the market, and a dearth, it seems to me, of time domain-capable programs. Why that should be so escapes me. An excellent example of a frequency domain analysis program is the shareware program ACIRAN.

If you're into recurrence relations as a model for DSP then it is simple to arrange each node of the digital processing scheme to have a column of its own. Delays are accommodated by using references to earlier rows in the formulae. It's rather more difficult to arrange for the precise kind of arithmetic which DSP chips tend to use; spreadsheets use floating point and don't know any binary!

The spreadsheet as database

As far as parts lists go, the first software application that might spring to mind as a candidate for keeping a parts system is a database. However, if a good spreadsheet has all of the database facilities we need (sorting, linking and extraction amongst them) and can perform technical calculations to boot, then why bother buying two different packages?

I've already described a sensible parts list arrangement in the section on documentation. Remember, if you've automated your parts list properly (and it is one of the best candidates for automation) then you are well poised to take advantage of that if you decide to go into full production. To take things a stage further: it's worth studying your spreadsheet and taking advantage of any extraction and reporting facilities to generate parts orders. Such a system ought to work sufficiently well until you are earning enough to indulge in a proper BOMP (bill of materials processing) system.

Spreadsheets are amazing things. If you can get yourself into the way of using them properly then you'll get a lot out of them.

CAD, CAE and graphics

For electronics people, CAD (computer aided design) covers traditional draughtsman's type CAD, the kind of 'CAD' we've discussed under spreadsheets above, and specialized CAD such as printed circuit layout and integrated circuit design which I suppose should really be called CAE (computer aided engineering). I don't propose to discuss integrated circuit design here, although it certainly has its own set of software tools.

Various degrees of integration are available with computer aided engineering. 'Point' or 'spot' solutions, as they are known, concentrate on a single aspect of design. At the other extreme, software is available which provides a complete service from capture of the circuit diagram through circuit simulation and on to design of the printed circuit boards themselves.

Basically, the circuit capture part of the system stores information in the form of a 'netlist' or list of connections from one component to another. Netlists can in turn be fed to the circuit simulator and the p.c.b. design part of the package. Systems such as this tend to be expensive.

Personally, I've always been disappointed by what passes for circuit symbols on most CAD systems of my acquaintance, especially the draughting or presentation graphics type systems. On those occasions when I've been using such packages, I've successfully made up my own circuit symbols; it's not hard to do but it all takes time.

I haven't come across any really neat shareware CAD. In the middle price range for commercial software, DRAFIX is a good bet. TurboCAD is inexpensive. Autosketch has recently come of age and is available for Windows now.

If you cannot afford CAD or presentation graphics, don't worry. You'll be reduced to drawing circuits by template, but some people find this easier anyway. As far as p.c.b. design goes, let me recommend Easytrax, the shareware version of Autotrax, especially as a first-time buy. If you have no computer and no printer, then p.c.b.s will need to be designed using tape; again, some people have never used anything else and swear by it.

A final word on Windows. I use Windows a lot. I use it because I can have many different things up on the screen at once, working in the way anyone works, i.e. looking at several things at once, not just one. There are a few die-hards out there who would not touch Windows with a barge pole, preferring instead to run one program at a time by typing in cryptic messages from DOS. If you do not want to type cryptic messages, and you have at least a '286 PC with a couple of Mb of memory, then Windows is highly recommended.

There are several important points about Windows: you only install printer and screen drivers, etc. once and all other Windows programs which are installed later run with the installed selections. You do not have to fiddle with these for every installation. The essential filing and editing commands on Windows programs are all very similar so you don't need to learn ten different ways of doing the basic things.

A sensible word processor comes free with Windows; if you don't need a spelling checker, it is as good as anything else. There is a calendar, a rather splendid calculator, a card file, a good file manager and so forth. Use these to organize your work. Finally, it's simple to set up a scheme which automatically loads the things you're working on at the time or which gives you easy access at the click of a mouse.

No, Microsoft did not pay me to say this. It is the original 'unsolicited testimonial'. Enough said.

Do the right thing when obtaining software. Obtain good title and obey the rules of copyright. If you can't contemplate the expense of the latest version, look out for older, surplus copies of the system you fancy; there are plenty about and it's not usually a problem to be a version or two behind. Hunt around for a good price; there's usually no need to pay the RRP and prices 'on the street' can be a lot lower.

When to upgrade to a new machine

There is certainly pressure from software, which tends to consume more computing power and space with each new version. There comes a point at which it serves no useful purpose to keep spending on an older machine; save your bucks and use them to leap a generation or two when the time comes.

For preference, when setting the new machine up, install software from scratch using the original media, learning from, and leaving behind, the old problems you had on the previous machine. The only problem you might have is untangling data and programs which are mixed up or relocated and retaining your customizations (see my previous comments).

Erase your old hard disk drive before selling your old machine; accidentally giving away software isn't really on.

Instruments

Without test instruments of some kind, you're poorly placed to test anything you've designed. Depending upon what you're doing, there are some instruments which are absolutely necessary, some which are nice to have and some which are merely convenient. They can range in price from a few pounds to many thousands of pounds for specialized gear.

It's surprising what can be done with the simplest instruments, if you know them well. I once had a boss who reckoned you just needed an AVO meter and a temperature sensitive finger, to do most service work at any rate. This is probably less true with modern electronics . . ..

Selecting, using and looking after test instruments is the subject of this section. I haven't gone into great detail about how to use these instruments; rather, I've concentrated on the things which are usually missed in the average text-book tutorial.

There is one activity which applies to all test instruments, that is, selecting the correct operating range. The right way to connect *any* instrument to a circuit under test is to select the correct function and range *before* connecting to the circuit. If in doubt, select a safe combination of range and function before connection. Usually, this means selecting a range which is known to be too large and then switching down a cog or two until a range is reached which shows the measured quantity to the best advantage.

Always take a quick look at the settings for any instrument before connecting up.

Many test instruments inject noise into, or load, the circuit under test. If connecting your scope to a circuit, for example, kills a nasty oscillation or starts one up, then it may be a sign that the scope is an inappropriate thing to use at this point in the circuit or (more important) that your circuit's reliability or stability are marginal.

Most bench gear is designed to be portable and is fairly robust. Just don't get it wet or drop it, unless it has been designed to take it.

Oscilloscopes

I suppose the oscilloscope (or 'scope' for short) is the archetypical engineer's 'thingummy' which the layman expects to see on every bench. If you hark back to science fiction television of the seventies there was always a scope embedded in a control panel somewhere on the spaceship's or submarine's bridge, with those fancy looping Lissajous patterns swishing back and forth.

The basic specification for a scope is its frequency response. 20MHz is adequate for a lot of purposes, certainly for audio work, although 100MHz is nice to have. You will but rarely come across a scope with only one channel; mostly, there will be two.

Delayed sweep is handy, as is the accompanying highlight control, as they allow you to 'home in' on any funny bits of the trace that merit a closer look. A storage scope is expensive but wonderful for catching transient and 'one off' events which otherwise defy examination. Delayed sweep, as such, is not needed on a storage scope as the sampling arrangements take care of that side of things.

With storage scopes, you usually get sensible timebase controls which correspond to the scope's maximum operating frequency, but very occasionally you'll get offered a scope with a weird timebase, so have a quick glance to make sure it's not one of the ultra-slow variety.

The ability to sum the channels or invert channel 2 is not that useful, unless you're working on power systems where earthing one side is not acceptable, in which case use both channels, summed with channel 2 inverted, to give a differential input. Please don't ever clip the earth clip of a standard oscilloscope to the neutral line of the mains. I've seen it done and it makes a lot of smoke and flames. You will be pleased to know that in this instance the scope survived longer than the operator's reputation.

The same comment applies to any instrument whose low-side input wire is not floating, i.e. it is not a true differential input. The trouble is that earth and neutral often have a few volts a.c. induced between them, and as they are low impedance sources a lot of current can flow . . ..

There will often be small knobs on a scope featuring a 'detent' (or 'click' if you will) at the 'cal' position. These allow a fine (but uncalibrated) adjustment of the timebase and the Y deflection controls. Quite honestly, I never use them. I suppose they're fine for photographers who want to fill the screen with a big trace. If your scope has these fitted then leave them in the calibrated position unless you're really going to use them. There's nothing more embarrassing than heaving a scope along to the departmental workshop to arrange for a repair, only to find that the little red knob is not in the cal position.

A scope is especially useful for analogue work. Digital signals can be treated a little differently. In fact, some digital signals are difficult to interpret on a scope; I'm thinking in particular of bus signals in a computer. On a scope, computer bus signals are just a hash of edges; 'grass', originally a term in use in the early days of radar, is the colloquial term for a mess of this kind, and very descriptive it is too.

There is, however, one useful feat you can perform using a scope on a computer bus; checking for shorted lines. If you get two *identically* weird signals you can be certain that the two corresponding bus lines are shorted together. To do anything more worthwhile with a computer bus you really need a logic analyser or at least a signature analyser.

Getting by without a scope is hard, but not impossible. I did without for some years. You have to live on your wits but it sharpens your skills in other directions! Expect to pay about £150 minimum for a scope, second-hand, or perhaps £100 if you're lucky. At the other end of the range, the sky's the limit.

Logic detection

A logic probe is not an expensive instrument as these things go. For digital work, it can even beat the scope flat. You see, a logic probe can catch the tiny pulses or 'glitches' that a scope might miss. Better yet, even if such a tiny pulse happens only once in every few seconds, you will still be able see it with the logic probe. It's difficult to see even a respectable pulse, of a few milliseconds' duration, using a scope if the pulse doesn't happen along often enough. For this reason, any logic probe you acquire *must* have a pulse stretch facility.

Often, the fact that there are changes of state of any kind is sufficient to tell you that a logic system is working. A logic probe works well at low speeds; its LEDs are easier to interpret than the straight line you get on a scope. The flickering of the LEDs and their relative brightness, which is related to the mark:space ratio of the signal under investigation, can tell you a lot about what's happening in the circuit. It's very handy to have a probe which will handle CMOS signals as well as TTL, by the way.

You can make up a quick, not very brilliant but adequate logic probe from a few gates and a monostable. Otherwise, expect to pick up a good one, new, for less than £20.

Sometimes on a computer or other logic system it's enough to know whether or not two signals are each in a particular state at a particular time. If you're really hurting to see that, then you could rig your own 'coincidence detector', consisting of a gate or two, on a scrap of stripboard. Check out the output of the detector using your scope or logic probe.

Multimeters

Broadly speaking we can class multimeters into the digital and analogue kinds, although some of the more modern digital multimeters (DMMs) have a moving bar display which moves in a similar way to the needle of an analogue type.

Analogue multimeters are not as robust as their digital cousins; they have a sensitive mechanical movement within and need to be protected from shock. They may also need to be used in a prone position - no, no, the instrument, not the operator! The movement is often designed to be used in that position and great inaccuracy may result from using the meter upright. The advantage of an analogue meter is that trends in the measured quantity are visible due to the behaviour of the needle, difficult to see on a digital instrument.

Switch analogue meters off when not in use. Unlike a DMM, an analogue meter will not necessarily drain its batteries if left switched on. The OFF position has another function. It shorts out the meter movement, protecting it to some extent, damping the swing of the needle as eddy currents flow round the short circuit.

There's not much you can do to a DMM (digital multimeter) if you're careful with the switch settings. I did see someone drop one once into a pool of sea-water, on an oil rig it was, which wasn't calculated to increase its chances of a ripe old age. I'm sure this is quite a rare event, but DMMs are so small they easily get pulled off things by the leads if they're left on an insecure perch.

If that ever happens to you, then get the batteries out and rinse the whole thing thoroughly in fresh water. Let it dry thoroughly before refitting the batteries. You might be lucky and revive it.

The same thing applies to any small, low voltage, battery-powered gear. I still have a calculator which I took to pieces and washed, inside and out, with warm soapy water. The keys had begun to stick, because the sugar in the coffee which I'd accidentally dipped it in had gotten sticky with age, I suppose. I let it dry thoroughly before re-assembly. It's still going strong twenty years later.

If you're going to use your multimeter outside a lot, there are units available which feature various degrees of resistance to water entry.

On the subject of using the switch settings correctly . . . checking the resistance between live and neutral or testing the current capability of the household mains supply doesn't do a DMM much good. My shiny new SOAR (well, it was shiny and new in those days) actually survived such an experience, but it still bears the scars in the form of a pair of rather wizened test leads. That'll teach me to lend it to people.

When checking resistance, tap the test leads together, before you apply the leads to the item under test, to see if you get the right reading or the beep for continuity. It's just a confidence thing for the DMMs, but the analogue meter may need you to make an adjustment before measuring resistance.

Multimeters are made in quantity, and some can be impressively cheap, but are likely to be inaccurate. The very cheapest analogue types are often cheaper than the cheapest digital types at about a fiver and £25 respectively (new). My own SOAR was rather less than £100. The best analogue multimeters are still, in my own opinion, the old AVOs. They can be had on the surplus market, but a good one is still worth money.

Counters, timers and frequency meters

Counters are simplicity itself in use, and are generally as accurate as a crystal oscillator timebase can make them - perhaps 30 parts per million, or ten times as accurate for the type in which the crystal is kept in a tiny oven.

The only general problem with counters is that of successful triggering. It's quite amazing to me that a scope can be made to trigger on anything from millivolts to mains voltage and yet the poor old counter seems to be denied a reliable triggering system. Some counters seem to be particularly poor on slow-moving waveforms.

One of the nicest counters I ever used, an old Bradley, had a pair of LEDs which indicated the state of the input. You could see what it was doing and adjust the offset control accordingly.

Some counters have a reciprocal facility for slow waveforms. Use of this facility prevents you having to use huge gate times of ten or even 100 seconds, waiting for ages while the count clocks up. The longer the gate time, the better the count resolution, but even ten seconds seems glacially slow sometimes. Note that the reciprocal facility is not the same as the period or time average setting, although that could be used to determine the frequency by taking the reciprocal using a calculator.

A decent counter might set you back about £100 new. Occasionally, you'll come across a DMM with a counter facility.

Power supplies

Clean, stable power is essential for your efforts. There's no point in compounding any problems you might be having with the circuit itself by using an unreliable or noisy supply. For an experimenter's bench power supply you should usually choose a linear supply.

You don't often need to go above 12V or so (15V is nice) and an ampere of current is adequate for most purposes. A variable supply is nice unless you're really intent on designing at one voltage only! For plant supplies, 24V, plus a little headroom if simulating a nicely charged 24V battery, is fine; say 30V.

Note that the middle connector on some supplies is an earth, connected to the chassis and mains earth, not a centre tap. Read the markings on the front panel to see what you've got. Supplies are usually deliberately left floating, i.e. the negative-most end is not connected to earth unless you choose to tie them together.

Switching supplies are not generally suitable for bench use for three reasons. First, one of the supplies (typically the 5V output) may need to be loaded to some extent before the others will come up to voltage, and they may not regulate well anyway on small loads. Second, they are not often adjustable. Third, they can be electrically very noisy, driving any sensitive analogue circuitry daft. On the other hand, if you're building a large 5V logic system, then by all means get one. They are not expensive if obtained from the surplus suppliers.

If you need split supplies and you've just got a single ender, the circuit of Figure 4.3 can give you a false middle rail at half the supply voltage.

Although you will not be able to sink or source more than a few tens of milliamps to such a centre tap, it will provide a bias point for further op-amp circuitry. Don't use the resistors on their own; the voltage at centre tap will vary if any varying currents are drawn from the junction of the divider resistors. The op-amp buffers the potential divider; we're not worried about the op-amp's few millivolts of offset. The actual resistor values are not critical. I've used circuits like this to provide a centre tap in systems powered from car and aircraft batteries. It works well.

The output of the circuit shown in Figure 4.3 is a *reference* voltage against which other voltages in the system are measured or compared. If heavy *supply* currents flow down the centre tap,

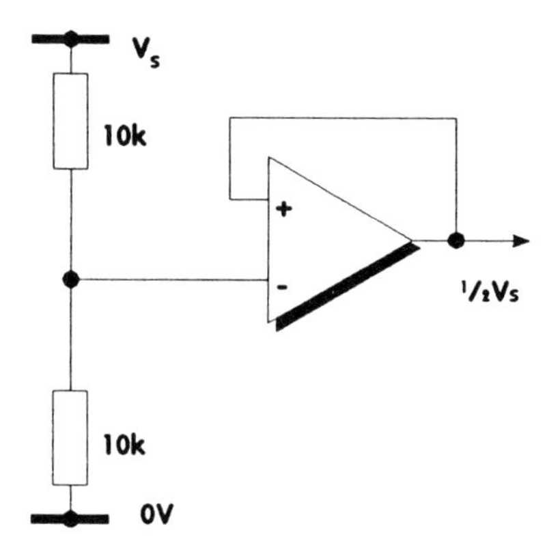

Figure 4.3: A split supply circuit

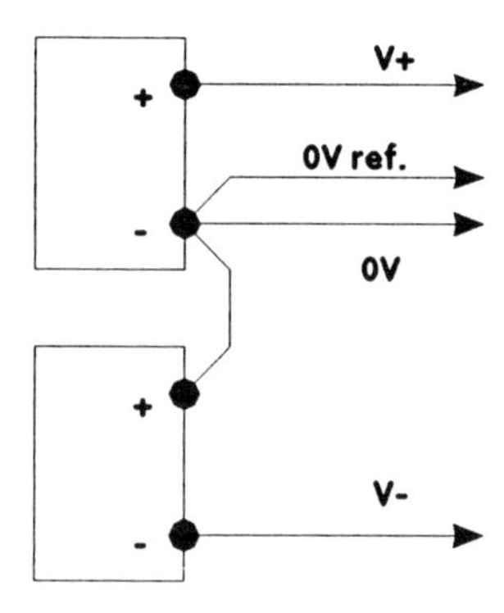

Figure 4.4: Using two supplies

use two supplies connected as shown in Figure 4.4, with the option of a separate reference wire as shown.

I'll be going into the implications of high currents in the zero volt rail in more detail later on.

Signal sources

I've grouped these together since they take so many forms and are often specialized for particular uses. Whether or not you seek to acquire (or build) one of these will depend heavily on what you're hoping to do. Signal sources are available to produce noise, pulses, sine, square and triangular/ramp waveforms or arbitrary waveforms. For those of a digital persuasion, bit sequence and word generators are available. Specialized generators produce signals with various forms of modulation intended for testing such things as television sets and communications gear.

These signals are all produced at various frequencies and amplitudes and with varying degrees of accompanying noise and distortion. The cleaner, more precise and more stable a generator you need, the more expensive it is going to be.

The commonest type is probably the ubiquitous, standard sine/square wave audio generator. The simplest is the crude 'signal injector', which has its uses in tracing the signal path, particularly through the likes of radio receivers and audio amplifiers.

The best advice I can give is for you to examine what you're doing and decide whether you have the signals you need already. The $64,000 question: is there a signal coming in from outside the system which requires to be simulated? Often, signals throughout a system will be derived from units upstream in the signal path. It may be that such a signal is a sufficient test without going to the expense of a signal generator. Signals from the environment, which are perhaps not conveniently obtained, may need to be generated.

Otherwise, you may have decided that you want to test the circuit using an input signal which it will not normally encounter, either to find out how the circuit behaves under extreme conditions or to confirm some aspect of its performance. You will certainly need a signal generator in order to do this.

To give a few simple examples: an electronic thermometer would not need the use of a signal generator in testing since the original signal, temperature, is freely available. Neither, on the face of it, would a radio telescope need the services of a signal generator; we would hope that the naturally occurring radiation would suffice for a good test. Circuits which are signal processors *per se,* i.e. audio filters, mixing desks and the like will inevitably require a signal generator for testing.

Expect to spend a good bit more than £50 even for a 'hobby' standard sine/square generator.

Dummy loads

As well as the usual collection of components which should be available to you, sometimes it's handy to have a set of monster resistors to load up any power supplies you're testing. As a variation on this theme, people who test radio transmitters often have loads called 'dummy aerials' which remove the need for an actual transmission while testing. There is inevitably some r.f. leakage, but most of the power is absorbed by the load.

Unless you're intending to develop a super power supply with amazing transient characteristics, there's no need to go out and buy a 'dynamic load', which is a rather sophisticated piece of kit which can be programmed to switch loads or vary them.

Analysers

Again I've classed these together. Spectrum analysers and logic analysers come into this bracket, although there are other, relatively obscure, kinds.

They are not found on every bench and you need quite special requirements, deep pockets, or both to be able to justify their acquisition. This situation is getting better and there are some scope add-ons which are worth investigating if you really need an analyser.

Use a logic analyser in those situations where you have a mass of digital data which needs to be presented simultaneously. You might, for instance, be tracking down a problem with spurious counting on a discrete logic system, or bad addressing on a microprocessor-based system.

Use a spectrum analyser in those situations where you're interested in the precise shape of a filter or where you're hunting around for some interference or distortion which has a particular frequency signature.

Consider hiring, too, if you really need something of this order for a week or two.

Component testing

Again, there is a whole class of instruments which are used for component testing. For prototyping, you may not be too interested in component testing. Where costly reworking is to be avoided, at the production stage, it may very well be worth investing in component testing, or inviting your supplier to do it for you.

You will test components for one of four reasons: first, to confirm that the component is within specification before committing it to assembly; second, to check out a suspected faulty component; third, in an attempt to select or match components with precise values; fourth, to find out the value of an unmarked or poorly marked component.

The usual check of a transistor using the diode function of a DMM or resistance setting of an analogue multimeter is inadequate as a test. I have known transistors pass this test but fail under the normal stresses of a working circuit. A transistor tester is such a simple device that it is hardly worth buying one; if required, it can be made up from a couple of resistors.

Where should you work?

The environment in which you work can have a tremendous effect on your success rate, as can access to the things you need and the frustration of not having things to hand. In as much as the place you work in is a 'tool', it is described in this chapter. It needs to be set up and used properly and kept reasonably tidy.

Physical environment

Your workplace should be neither too cold nor too hot. 20°C (70°F) is supposed to be a good 'sitting temperature', but this will depend on other things like air movement and humidity.

Lighting is important. You might be able to see to read, and it's wonderful how your eyes can accommodate, but for seeing fine detail you need good light. Spot lighting is fine, but remember that you need to see around the rest of your work space too; a single anglepoise lamp is not adequate. Soft, shadowless lighting is great. Highly directional lighting is not very useful. It's unnecessarily frustrating to not be able to see properly, and the fix is simple: use adequate lighting.

If you're a computer user then lighting takes on a whole other dimension. Staring at a piece of glass in which there is a reflected image of the overhead light or your own nice white shirt is galling. Ideally, you should site your computer at right angles to the prevailing direction of any strong light, so that it shines neither into your face, nor onto the computer screen.

Music might help the creative juices if you're that way inclined but remember not to inflict it on anyone else, particularly not a customer.

Make yourself comfortable. Don't get too cramped. As I said before, find an excuse to move around now and again.

Bench layout

The ideal place for the instruments is above the bench. Make sure that any shelf you have the instruments perched on is up to the job. Don't have any such shelf too high; after all, you want to be able to reach the controls. The other thing about having instruments far away is that it makes the test leads longer, bad practice as it means that interference is the more easily picked up.

Benches are available with built-in instruments. Personally, I like to be able to leave things in position but to lug them about or change them if I want. If you have no instrument shelf, push the instruments to the back of the bench or to one end of it.

Your bench should have plenty of power sockets. Site them between bench level and the shelf which on which the instruments sit. You might need four even just for small jobs; scope, power supply, soldering iron, signal generator. Eight is better; you almost can't have too many.

Soldering irons and test leads don't mix. Find a place for the soldering iron out of the way where stray cables and leads will tend not to drape themselves gracefully across it. The other thing to watch with test leads is that you don't pull them off the shelf. You might decide to keep small instruments, i.e. the DMM, down on the bench where they can't fall much further.

Lastly, try and keep hacking, cutting, drilling, sawing and filing away from assembly, even at this stage of the game. Apart from the obvious issues of keeping metal filings away from circuit boards, you're begging for an accident with a crushed board as some extra heavy tool is laid down in the wrong place.

Storage

Everything in its place is a good motto. Therefore: everything must have a place. There's no need to spend a lot. Recently I had the good fortune to be offered an old chest of drawers, which swallowed my entire stock of components and junk.

The humble cardboard box should not be despised as a storage medium. I knew a bloke who operated out of his kitchen; he kept his entire stock of bits and pieces in labelled boxes on shelves high above the kitchen table. He knew where he could lay his hands on any component within a minute or two; all his partly completed work fitted neatly in there too.

If you want to get a little fancier than cardboard boxes, or if you feel that your official visitors might not be impressed by an old chest of drawers, then the multi-coloured plastic trays which stack neatly or fasten onto a louvred panel are a good bet. They're not desperately expensive either. They're available in anti-static versions too. But you don't need those because you always keep your chips in their anti-static packing until you need them, don't you? Don't you??

Not being able to find a component you're sure you've got is very frustrating. I'm sure I had an old 74LS02 in here somewhere, he cried, digging in the dusty depths of an old sandwich box. On the other hand, you can be pleasantly surprised. My supplier was out of 32kHz watch crystals the other week and just as I was getting ready to lay that project aside I came across one I didn't remember having. *C'est la vie.*

Why not save any self-seal plastic bags and chip carriers; in modest quantities they'll come in handy. If you're into photography, the plastic bottles which the film comes in are often handy for keeping small parts in. If you've space, keep as much of the bubble wrap and perhaps some of the polystyrene 'peanuts' salvaged from your incoming parcels as you can. This stuff comes in handy if you want to send things away yourself, although you'll need to be a bit more formal with your packaging if you go into production.

Banker's boxes are useful for magazines, data sheets and those silly thin catalogues that won't stand up for themselves on the shelf like *real* books.

If it's a production facility you're after, then you must be more careful about siting, about storage and about providing adequate space, even if it's still only you in there. One can tolerate a little awkwardness in the breadboarding environment which would be unworkable in a factory kind of setting; I've written about that further in the chapters on production.

So how much room do you really need? Work spreads to fill the area available for its completion - the physical corollary, I suppose, to Parkinson's Law. The space in my little office which is devoted solely to electronics is just about one square metre. In all honesty, the floor gets used as well sometimes! I would like more space and I'm not as tidy as I ought to be, but I still manage to produce some nice kit. It must be OK since people keep coming back for more.

It is amazing how some people make use of space. I remember reading, many years ago now, an article in a magazine about using a spare broom cupboard as a minimalist workshop. A full sized shelf at about desk height formed the main work area; above this, instruments could be racked on other shelves, the probes dangling within reach.

At the rear, there was a built-in experimenter's power supply. There were drawers beneath for components and tools; large items sat on the floor or went up into the gods. When it was time for tea, the stool got shoved under the bench and you slammed the door on the whole kit and caboodle.

If you're contemplating such a minimalist workshop, just remember to switch off the soldering iron before closing the door . . ..

5: Design Principles and Techniques

This chapter covers the uses of partitioning, algebra and Boolean algebra, all useful techniques for 'modelling' your circuit or system. When you come to build your project, whatever it may be, it will stand a better chance of working first time because you've modelled it properly.

Since the division between hardware and software is a problem in partitioning, the important questions of where to draw the line are addressed here too.

Also described here are how to use calculations, approximations and rules of thumb.

Partitioning: breaking your design up

System diagrams and block diagrams are the first exercise in partitioning. When we prepare such a diagram, we are attempting to make a first guess at the internal workings of the system. We are splitting the system into a number of smaller subsystems.

The reasons for doing this is that it breaks the design up into more easily manageable parts. Just as likely, it will also result in a physical separation of certain subsystems. For instance, it is simpler to design the r.f. parts of a radio receiver on their own rather than trying to integrate the audio sections at the same time. Eventually, once the size and nature of the parts has become clearer, the decision might be made to keep the two physically separate on two boards in the same cabinet.

The split of our hypothetical radio receiver into two parts, the r.f. and a.f. parts, is called logical partitioning. If we decide to keep the two on separate boards or in separate cabinets, we will be engaging in physical partitioning.

Exactly how we partition a system will depend upon our own experience and the availability of further information. Even at this stage, no two person's solutions will be identical. You are not 'wrong' simply because your efforts are not identical to those of your colleagues. You will not be wrong if you disagree in detail about how I've partitioned my own examples; nor will I.

Often, we'll do much of the logical partitioning first and then, when we're satisfied that we've reduced the system to a suitable level of complexity, we will start to think about physical partitioning. On the other hand, some aspect of physical partitioning may be inherent in the specification and we'll doubtless incorporate that into the plan right away.

There is an art in deciding how big the blocks must be and where to split. Sometimes it's difficult to know where to start. A good point to start from is to sketch a single box which represents the 'system' as a whole, along with the inputs and outputs. That defines the problem and concentrates the mind on the question: What has to happen inside the box in order to get the behaviour we want?

I'm not sure who I'm paraphrasing here, but I'm sure someone must have said it before at some time or other, so it's in quotes:

'*A division in a plan, device or system should be at that point which minimizes the complexity of the interface*'.

Incidentally, read the word 'interface' in the quotation above in it's most general sense, rather than the special electronics/ computing sense.

Most successful designs employ this philosophy of least complexity. Any form of technological design benefits from the application of the rule. More often than not it happens by chance or because the designer has a flair for it or she 'feels that it's right'. You have the advantage of having it spelled out for you.

By drawing up a practical and sound system or block diagram, we are obeying all the best rules of cognitive psychology - minimizing the complexity of the interface between a system and its designers.

What is complexity, then? If something is made up of many interconnected parts, then it is said to be complex.

From the electronics point of view, more wiring always means greater complexity. Minimizing the number of wires which run from one board to another, from one cabinet to another or from the control panel to a board are all good things to think about. In electronic systems, the number of wires we need to run between parts of the circuit is often a good guide to complexity. The fewer wires which run from one part of a system to another or (as far as documentation is concerned) from one page of a circuit diagram to another, the better off we are in terms of reducing complexity.

Very large or very tiny amounts of current and voltage may imply greater complexity in a slightly different way, as might high power handling requirements. Due to the ways in which large amounts of power or very tiny signals need to be handled, the circuitry itself may need to be more complex.

Where lots of information must pass from one part to another, keep those parts close together, if possible. Look out for groups of functions which relate to each other and see about keeping them in the same place.

For the moment, we're not interested in what goes into the blocks of a block diagram. We're interested merely in the relationships between blocks, that is how they are connected together.

Transport designers make use of natural divisions between the power plant, the vehicle body and the suspension/road wheels. They must all fit together, but once the relationship between the parts, the 'fitting together', has been specified, the parts can all be purchased or designed and built separately.

In the same way, there are natural divisions in an electronic system which we can take advantage of. Perhaps the most natural distinction an experienced engineer makes is between the system itself and its power supply. Further natural breaks come between, say, the r.f. sections and audio sections of a radio, between the transmit and receive parts of a transceiver and between pre- and main amplifiers in a stereo amplifier.

Where a different signal regime takes over is a good place to put a partition, e.g. wherever we've demodulated the r.f., digitized the analogue or applied gain. Convenient places at which to access signals or inject signals are also convenient places to put a partition.

There is no limit to the subdivision which we can indulge in, until we hit component level. Components are the atoms of our electronics universe. We carry on partitioning until we get to component level, recording the results of our partitioning on several different documents, block diagram, physical layout, circuit diagram, so that we have appropriate information available at the time its needed.

Often we'll buy in a self-contained unit of someone else's design. Treat this as you would any other component. Power supplies, filters and single-board microcomputers are examples of this. On the block diagram, just leave them as a single block. We're committed to using them in the form in which they are given us; trying to divide them up is not sensible.

Software engineers often refer to a process like this as 'stepwise refinement' - a nice phrase.

I once came across a rather extreme example of bad partitioning. Some person had run a couple of wires from one board to another just to use a spare inverter gate. Please don't do that, it's horrid. Yuk!

Partitioning hardware and software

This represents a major and important division in any system which is to incorporate a microprocessor or which is to be included within a system along with some form of computing power. Perhaps you're designing a plug-in board or an external serial device, or perhaps you just need the odd bit or two of control for a heater, motor or solenoid. The kinds of projects which incorporate computing power are increasingly common.

Throughout this section I have used the term 'software', although once installed in read-only-memory (ROM) it should, strictly speaking, be called 'firmware'. Install your software as firmware in ROM if you are building dedicated controllers which do not feature discs or other storage media, which need to function independently or which have a need for physical ruggedness.

Occasionally, you may be designing for some system in which you have no control over the software, since your design must work with existing software. In that special case, you have no option but to partition in accordance with what you know about the existing system. Otherwise, there are some simple differences and trade-offs between hardware and software over which it is worth spending a paragraph or two.

First, software is slow. All things being equal, software is inevitably slower than hardware. Any given software instruction takes a few hardware cycles to complete. Also, hardware is inherently parallel; where our processor might have taken 50nS to complete a logical AND instruction, the hardware might actually have made several such decisions using several separate AND gates in the course of 10nS or so.

The problem is exacerbated when using high-level languages which inevitably are not as efficient in their use of the memory and processor resources, either space-wise or speed-wise, as assembler. Some come pretty close, C being a good example, but compilers cannot spot all of the short-cuts of which hand-crafted assembly language can take advantage. Even worse are the interpreted languages, such as most BASICs; only use these if you are designing a snail checker or glacier tester. Where speed is important, a hardware solution is probably better.

Not only is software slow, but it is also ill-timed. This is a particular problem where interrupts interfere with the normal running of the program (I have more to say on interrupts in Chapter 9). Interrupts are a feature of any modern computer, large or small, especially those which have resident software which needs to be alert for input from keyboards and other peripheral devices.

Where timing is critical, you would be well advised not to trust to software timing loops at all, since these will expand to accommodate the time used in servicing interrupts. Either use a polled timer to get a true estimate of the elapsed time, or better yet arrange to prime a timer chip with a delay and then set it off, letting the peripheral hardware romp along without further interference from the processor until the time comes to call a halt.

It takes a lot of processor time to continually poll a timer; you can maximize the efficiency of your design by using hardware which does not need to be continually serviced in this way. Perhaps an interrupt could be raised, once the counter has run down, to inform the processor that a certain time has gone by. On the other hand, if that's *all* you're processor has to do . . ..

It is easy to make software complex. This might be seen as a bad thing, but in fact you need something complex to control a complex process. A particularly good example of this is in loops. Hardware which cycles around an indefinite number of times, successively subtracting and testing a value then progressing onto the next phase of operation, is hard to build. Far better to incorporate a loop in a microprocessor program.

Microprocessors are good at arithmetic and decision making, so any situation in which the choice of one of several actions might depend upon the outcome, some fairly complex calculation is a candidate for the use of a microprocessor.

Third, software can be changed without recourse to the soldering iron, unless you've soldered your ROM straight to the board! This has advantages which need not be explained at length.

Taking all this into account, it should be possible for you to decide upon which parts of your system need to be implemented using software, if any.

Normally, simple actions can be taken in hardware. However, where you must have a microprocessor for other good and pressing reasons, then there's no problem with looking around for other suitable functions which can be incorporated in software, thus maximizing the benefit of having the microprocessor there. So be on the lookout for other things which can be palmed off on the processor, if you're seriously thinking of having one, without necessarily overloading it.

Deferred design

Deferred design relates to putting off the commitment to finalise a design, preferably so that someone else, not necessarily the ultimate end user, can program the device to *specialize* it for a particular task.

Any design has a deferred element in it somewhere if you look hard enough. Every device has some feature which is changeable in use, or else the device can be used for some purpose other than that for which it was intended. I suppose the best example of a deferred design is the microprocessor. The microprocessor is the ultimate programmable/alterable gadget, hence its popularity, with engineers at least!

It may sound as if deferred design is intended as a kind of cop-out for the designer. Really, it's a kind of forward-thinking partitioning, a partitioning in time rather than space. What we're attempting with deferred design is to anticipate possible future needs and provide a neatly packaged subsystem which precludes the need to design similar things from scratch in the future.

It takes very little effort to design something with proper interfaces which can be used in a variety of situations. With a good deferred design, unthought-of situations arise in which the existing design can play a role, just by customizing and plugging it in. If you have been careful to design a robust interface for each subsystem, then it will be easy to reuse.

As well as designing in a good electrical interface, deferred design imposes a little bit of organizational overhead: we need an interface, too, for the people who are going to customize it. Hence, a user guide or similar, giving details of how to customize the various options, is needed.

Deferred design does good things for the robustness of a design, by imposing the need for a sensible interface mechanism, but may or may not have a beneficial effect on economies of scale during production. The extra cost of providing a general purpose interface which may not be needed for many applications may impose unwarranted costs. But if you're putting together a number of customized or specialized systems using building blocks, then a wholesale deferred design approach is the thing for you.

It is a bad sign if you have to redesign basic building blocks for special purposes. On the other hand, you need not use deferred design on subsystems which you feel have no future in other products and other places.

An ideal example of deferred design for our purposes might be a filter circuit. We design a good filter which has the capability for being customized for a number of different uses, then we either recall it as a p.c.b. macro from computer disk, ready to be incorporated as a *logical* partition of a larger system, or we build a stock of actual *physical* objects to plug into our new design.

Arithmetic and algebra

Not only is algebra useful in itself, but it also forms a basic part of getting spreadsheet functions to work. As I said previously, the only venture into mathematics for this book is going to be some simple algebra. Algebra is arithmetic with symbols, if you like. It's amazing what you can do and how far you can get if you are conversant with algebra and are not afraid to use it.

Algebra is one of the essential thinking/modelling techniques; it's well worth getting familiar with it. Most readers *will* be familiar with it, but if you are not, or would like a summary, then read on.

The most universal and useful piece of algebra you'll come across in electronics is the equation $V = IR$. The symbols V, I and R represent voltage, current and resistance, which are measured in units of volts, amps and ohms respectively. So far so good.

Now some people have trouble with symbols. This doesn't mean that such people are flawed in some way, just that everything they've ever done in the way of arithmetic has boiled down to a number at the end of the day and they're not used to the idea of symbols. A symbol just stands for a quantity whose value is (presently) unknown or a quantity which we know about but can't conveniently commit ourselves to yet.

We use symbols because they make the equation easier to read, because we don't yet know what value to give the symbol, or because the system we are modelling can be succinctly described in all its important aspects by such an equation. An equation is a model (there's that word again) of some aspect of a situation or the behaviour of some component.

Sometimes a symbol is called a 'variable'. This isn't strictly good form, since some symbols can be 'constants'. The speed of light is a good example; it is a constant which is often represented by the symbol c.

For example, with a little experience we can tell, at a glance, that $V = IR$ refers to voltages, currents and resistances, whereas $2 \times 3 = 6$ could refer to bags of peanuts for all we know. Add to this the fact that we can append subscripts to the symbols (i.e. V_{out}) and we have a powerful ability to build models which represent any quantity by name.

Because it makes the equation easier to manipulate, it's a good idea to leave the symbols in until the very end before substituting real numbers and doing the actual calculation itself. By substituting numbers too soon, we lose track of what the equation actually means; the description is lost. The novice's first reaction is to grab the calculator; please don't do that.

The equals sign is the most important part of the equation, really. An equation gets its name from the fact that it contains an '=' sign. We must take great pains to make sure that it always remains true. In other words, if we do something which makes the two sides of the equation unequal, then we've made a false or misleading statement. The golden rule is, always do the same thing to both sides of the equation.

$V = IR$ tells us that simply multiplying a resistance by a current gives a value for voltage. This is *always true*. There are no known exceptions; that's why it's called Ohm's Law and not Ohm's Postulate or Ohm's Hypothesis.

But say we know the voltage which exists, we know the current we want to flow and we are trying to find a suitable resistor value? Then we need to re-arrange the equation to make the 'unknown', R, appear on its own on one side of the equation or the other.

Remember that we must do the same thing to both sides of an equation. So if we divide both sides by I we get $V/I = IR/I$. I now cancels on the right to give $V/I = R$. We now know that it's sufficient to divide the voltage by the current to get at the resistance. It's a cinch; great.

Any kind of manipulation like this is allowed provided we obey the rule. The trick is in deciding what to multiply or divide by (or add or subtract as the case may be). To do this, find some operation which, when applied to both sides, results in one side becoming simpler or more like the unknown variable or symbol on its own.

One thing which is always assumed but never seems to be pointed out is that the quantities being manipulated must relate to the same component or other entity. You find confused people trying to multiply the resistance of a particular resistor by the current flowing somewhere else in the circuit! Come along now; what nonsense!

Equations of the form $V = IR$, which relate three things together in a multiplicative sense, are very common indeed. Power, voltage and current are similarly related, as are charge, current and time.

Dimensional analysis, that is, taking into account the units in which quantities are expressed, can be very useful as a check on the correctness of an equation. Basically, we cannot add quantities which do not have the same units. One gallon plus five inches obviously does not make sense. Neither does 400 ohms less 20 milliamps. If ever an equation of yours threatens to produce results like that, then scrutinize it very carefully; it contains an error, guaranteed. Be aware of the units in which the unknown is expressed, then make sure that the other side complies.

Division and multiplication change units, whereas addition and subtraction do not. Adding miles and hours is nonsense, but dividing miles by hours produces those units of common experience, miles per hour.

While we're on the subject of units, please make sure that you quote the units for the results of any calculations you do. If the result is in, say, ohms, then say so.

You will also come across numbers with no units. These are often called 'ratios'. Gain is such a ratio.

Take for instance an amplifier whose output voltage changes by ten volts whenever the input voltage changes by half a volt. The expression for the behaviour of such an amplifier (its 'transfer function' if you like) looks something like 10V/0.5V. Numerically, the expression evaluates to 20. But the units cancel to give us no units! We have ourselves a ratio. Don't ever use the word 'ratio' to mean any number which has associated units.

As for actual numbers, facility with the manipulation of exponents is needful, as is familiarity with the decimal point and its working. It may seem like kindergarten, but when converting between numbers expressed in the natural way and numbers with a power of ten, I use the old trick of making the decimal point jump the digits until it lands where I want, counting the number of jumps it takes. I then say ten-to-the-jumps. It works, so why not use it?

What foxes some people using this system is the implied decimal point at the end of a whole number (integer). We can't make the decimal point jump; there isn't one, they say. There is; just dot it in after the last digit.

In actual fact, if you use very small and very large numbers often you'll quickly develop a facility for them and you'll wonder what all the fuss was about.

We will be looking at the business of 'significant figures' when it is time to choose real component values.

As for the more advanced stuff; well, if you insist on designing filters from scratch, then perhaps complex number theory is needed. Calculus is applicable to some areas.

Personally, in all my eighteen years of buzzing around doing real electronics I can count on the fingers of one hand the occasions when I needed some advanced math (apart from those times when I was actually involved in teaching it or studying it for its own sake). Knowing about advanced math can still be advantageous; you may have a speciality which requires the use of some advanced technique and it certainly enhances your understanding of the simpler things. When you *do* need advanced math, you need it badly.

I enjoy doing beautiful math, it gives me a cosy feeling of completeness and security when I can describe something succinctly in a mathematical way. But for the ordinary things in life, forget it.

Critical and non-critical values: rules of thumb

Component values are more critical in analogue circuits than they are in digital circuits. Analogue circuits rely totally on the precision of individual component values to give the precision required in the processing of any signal. Precise component values tend to be needed, in particular, in those instrumentation circuits where some signal is being generated or being measured with some degree of accuracy.

Not all component values in an analogue circuit are critical. Those situations in which we can use a resistor of wide tolerance are explored in the examples later.

Digital operation generally is insensitive to wide tolerances in component value, most passive components being for non-critical uses such as decoupling, pulse generation and pull-up or termination. Digital signals themselves are not subject to corruption within wide margins and the actual voltage can vary, within limits, without the logic state being corrupted.

Calculation gives a precise value for the component values we need. Such a value can, however, be spuriously precise. Take, for example, the value of a resistor which is to pass 4.5mA when it has 8V across it. Faced with this calculation, your trusty Casio will say 1.777777778 03.

Well there's no way that you're ever going to get a resistor of that value and that tolerance. The nearest standard resistance, 1k8Ω, is a little more than 1% too large. If you land in this situation (inevitable, really) then there are three ways of overcoming the need for odd component values.

First, just use the nearest standard value. Check out the performance of the resulting circuit, both in theory and then in practice. Often, it's possible to alter another value somewhere else in the circuit to compensate for any error.

Second, make your own component up from two standard values in series (parallel for capacitors). The calculations for this operation are a candidate for a spreadsheet if your macro programming is up to scratch, or perhaps you might knock up a little something in BASIC. Two components combined in this fashion will have a temperature coefficient and precision no worse, on a percentage basis, than that of the individuals.

There's no need to try getting any closer to the desired value than the tolerance of the individuals. Note that this second option is not the same as 'adjust on test', a technique of selecting a component at production time. Avoid doing that if possible.

Third, add adjustments. The proper uses of adjustment are tackled in Chapter 9.

The number of situations in which a component value is not critical can be made greater by using circuits which are either adjustable or inherently non-critical in some way.

One of the rules of thumb which I employ constantly is the one about LEDs having a forward voltage, at the rated current, of 2V. If you think about it, this assumption becomes less valid at low supply voltages. I always aim at about 20mA of current, choosing a standard resistor value to give me a little less if that's necessary.

Given a little practice, you'll almost certainly develop your own ways of quickly assessing your favourite circuit fragments. There are no easy rules about this kind of thing, except to consider what might happen if the value of the component you specify really does change by the full amount allowed in the tolerance.

Tracking currents through a network

Tracking currents through a circuit is an important aid in design. Remember, electronics is all about controlling currents. For most circuits, if we can estimate the flows of currents then we probably have a good working knowledge of that circuit. The technique is particularly applicable to analogue circuitry at d.c.

Tracking currents in the imagination is even more important than the ability to track them with an instrument. Kirchoff reckoned that all the currents at a node (i.e. where things join up) sum to zero. Bear in mind, when trying this out, that any currents flowing away from the node in question are negative by convention. See Figure 5.1; the currents flowing are modelled by the equation

$$I_1 + I_2 - I_3 - I_4 = 0$$

A fact which, amazingly, passes a lot of people by is that the current coming out of a component is the same as the current going in. It doesn't matter what that component is, the current at one end is the same as the current at the other. This is a natural consequence of Kirchoff's Law; the component acts as a node in this case. Three-terminal devices obey the rule too; the sum of the collector and base currents in a bipolar transistor, for example, equal the emitter current. After all, current does not disappear into thin air, and where else could it possibly go? Figure 5.2 demonstrates this graphically.

The bias currents I_{inv} and I_{non} to the op-amp are very small and may flow in either direction depending on the op-amp's type and operating conditions. Even so, the rule still applies. In particular, remember that the op-amp's output current comes from the power supply currents I_+ and I_-.

Even in capacitors, which you might think of as soaking up current, the current flowing in at one end 'chases current out of the other', so to speak, balancing the charges in an equal and opposite manner. If one end of a capacitor is not connected to anything, the charges cannot flow into or out of that end and the capacitor will not charge up and develop a voltage across itself.

There are other ways of analysing circuits, more mathematically inclined, which are very useful. I've presented current tracking here as the most widely ignored, yet simplest and potentially most useful way of gaining understanding of circuit operation.

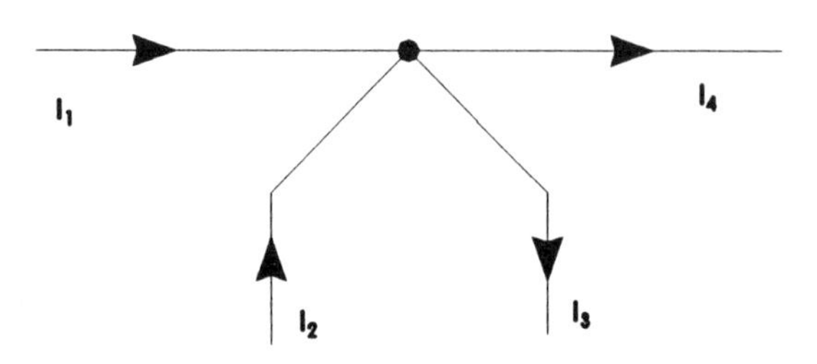

Figure 5.1: Currents flowing into and out of a node

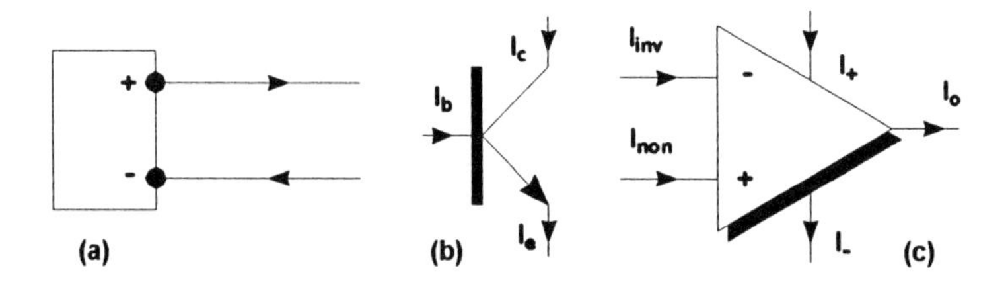

Figure 5.2: Current flow through a) PSU; b) transistor; c) op-amp

Boolean algebra and other digital tricks

When designing logic systems, there is another algebra apart from the familiar algebra of ordinary arithmetic which can be usefully employed. This Boolean algebra, named after its inventor, has variables which may only take on one of two values, a zero or a one.

It also dispenses with the usual addition, multiplication, and so forth and uses the logical operators AND, OR, EXOR and NOT, traditionally represented by the symbols ., +, ⊕ and a bar over the variable name respectively. To aid in legibility, you could use the ampersand & for AND and the tilde ~ for NOT.

In the same way that multiplicative operations take precedence over summative operations in ordinary algebra, so also does AND take precedence over OR and EXOR in Boolean algebra. Just as in ordinary algebra, use parentheses () to alter the order of evaluation or to remove any potential ambiguity.

Thus an equation relating the two Boolean variables A and B to a result C using a NAND (NOT AND) gate would be:

```
C = ~(A & B)
```

When faced with a large set of gates, such as the decoding scheme for a microprocessor, it is often easier to reduce its behaviour to a Boolean equation than to tackle it by working through it in the usual way, muttering incantations such as 'if that one goes high, and that stays low . . .'.

I remember using Boolean algebra in an effort to sort out the innards of a fire and gas warning/shutdown system on an ageing oil platform in the North Sea. Although the logic was worked by relays rather than chips, Boolean algebra was most appropriate and served us well.

You will inevitably come across truth tables. Strictly speaking, some of the tables known as 'truth tables' should really be called 'behaviour tables' since they refer to pulses, edges and stored bits as well as logic states. Behaviour tables are given a good work over in the Chapter 6.

Incidentally, AND is often referred to as the product of the inputs and the result of ORing as the sum.

For now, take a look at Table 5.1 and Figure 5.3, which show the truth table and circuits for the equation above. Either of the circuits has the same logic function, but one is implemented using a single NAND gate and the other is implemented using a separate AND and NOT gates.

A	B	A & B	~(A & B)
0	0	0	1
0	1	0	1
1	0	0	1
1	1	1	0

Table 5.1: Truth table for AND and NAND gates

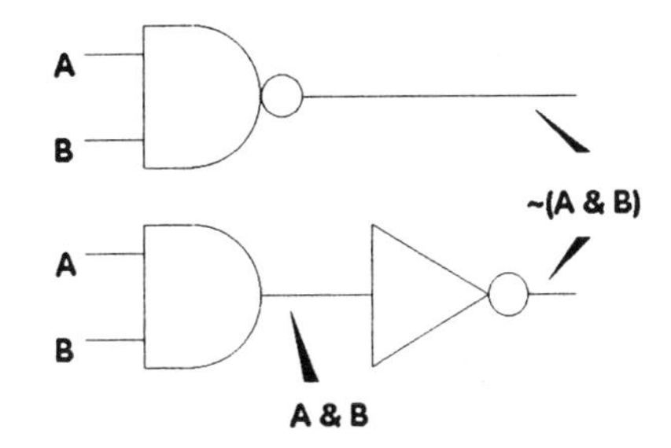

Figure 5.3: NAND gate implementations

You should note that the inversion is applied to the product of A and B and not to the individual inputs. ~(A & B) is not the same thing as (~A) & (~B).

There are other digital tricks around, the most well known is perhaps the Karnaugh map, a way of minimizing logic systems once the behaviour has been determined in terms of inputs and outputs.

Boolean algebra on its own, however, is a good bet to get you out of most kinds of digital trouble.

6: The Behaviour of Real Components

Components are the nuts and bolts of electronics. In this chapter I've described the components themselves and the implications which the underlying technology has for their individual behaviour. The next chapter provides hints on how components can be combined into working circuits.

The actual electrical element in many electronic components is physically very small. Most of the bulk of a resistor or an integrated circuit is to do with mechanical support of the circuit elements themselves, getting heat out of the component and getting connections into it. Inductors and capacitors are different; their bulk is mainly due to the quantity of metal and insulation they need to do their jobs properly.

Passive components

All passive components can be classified as R, C or L (resistance, capacitance or inductance). Despite this broad classification, components are not perfect; they tend to have tiny 'parasitic' amounts of other things in them, as we shall see below. For instance a resistor, unless specially made, usually has a certain amount of inductance.

Resistors

Resistors can be loosely classified according to their construction. There are really three ways of making a resistor.

First, wind resistance wire onto an insulating core. These kinds tend to be for either precision use or for high power handling, and tend to have a large parasitic inductance. This is avoided in the 'non-inductive' types by winding half the wire backwards to give as many turns in the opposite direction.

Second, put a film of something conductive or resistive on an insulating core. Metal film, metal glaze and carbon film types are all like this. They have less inductance than the wirewound types, although the inductance can still be substantial if the resistors are further processed or 'trimmed' by cutting the film into a helical pattern.

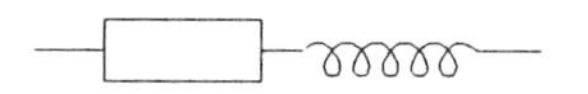

Figure 6.1: Equivalent circuit for a resistor with inductance

Third, use a block of some partly conductive material. The really old carbon rod type were made this way; these had poor characteristics in many ways and are rarely seen nowadays.

For low frequency use, the parasitic inductance is immaterial. Figure 6.1 shows the equivalent 'lumped parameter' circuit for a resistor with parasitic inductance.

Tolerances are an important feature of all components. One percent seems to be the usual for resistors nowadays; there is very little to be saved, except when buying large quantities, in going for the 5% or 2% kind. Going to 0.1% is expensive. There are resistors of tighter tolerances but they are *very* expensive.

If you have a resistor whose nominal value is 100Ω and which claims 1% tolerance then its value is almost certain to lie within the extremes of 99Ω to 101Ω.

Resistors change their values as the temperature changes. This change need not always be a higher resistance for a higher temperature and will vary between individual resistors. Think about using resistors with a low 'tempco' if you are designing precise analogue circuits which must operate over a wide range of temperatures.

Resistors dissipate heat. The temperature rise due to heat dissipation will add itself to any ambient temperature fluctuations as a cause of change in resistance value. The rate of heat dissipation (power) can be calculated, most simply, by multiplying the voltage and current together. Where your resistor is going to dissipate appreciable power, rate it accordingly.

The other possible cause for grief with a resistor is the noise generated. Some of this noise is a function of the absolute temperature (in Kelvins) and the resistance value, and is a physical fact of life which can only be alleviated by keeping resistance values small or by reducing the temperature. Noise due to the construction of the resistor or faults within it is known as 'excess noise' and is present in addition to the thermal noise.

Fortunately, the noise figure of any sensible resistor is vanishingly small and can for most purposes be disregarded.

A resistor is used to 'define currents and voltages'. That sounds a bit vague, doesn't it? Let's explain. You probably have a known voltage between two points in a circuit and you wish to limit the current that flows between those two points to some well defined value, so you use a resistor between those two points. Conversely, you may wish a particular current, flowing between two points in a circuit, to generate a certain voltage. Again, use a resistor.

Resistor values vary from fractions of an ohm to many millions of ohms. The most useful values range from ten ohms to a megohm (1MΩ), with a thousand (1kΩ) to a hundred thousand (100kΩ) being most commonly represented.

The reason for this clustering is that with this order of resistance we are dealing with sensible voltages and currents, tens of millivolts to tens of volts and hundreds of microamps to tens of milliamps. With very small resistances you're dealing either with high currents or tiny voltages (remember $V = IR$?) and the resistance of any associated wiring, soldered joint or switch contact could be an appreciable fraction of any designed resistance, making special precautions necessary.

At the other extreme, I have come across resistors with values in the thousands of megohms; they are specially constructed with glass envelopes. You certainly can't solder them to an ordinary p.c.b. The resistance of the p.c.b. might be comparable to that of the resistor itself (due to leakage currents through surface contamination). Touch such a resistor and its resistance changes due to your fingerprints. Electrostatic shielding may be required when employing such resistors too.

For ordinary run of the mill circuitry, try to specify resistances in the tens or hundreds of ohms to the hundreds of kilohms where possible. If you like, and provided it doesn't compromise circuit operation, minimize power consumption by using the largest sensible resistor values. You've usually got a choice and it means that you're working with sensible current and voltage values. For battery-powered equipment you'll want to pay more attention to this power-saving aspect.

For micro-power circuitry, the extreme low-power gear, one of the techniques you'll use will be to take resistance values up as far as practical, since you'll want to keep power consumption as low as possible. The values of all other components are subordinate in that case.

One interesting use for larger power resistors is as small heaters. Their dissipation, usually an unwanted feature, can be put to good use. Some resistors will mount on a heatsink, increasing their maximum allowable dissipation.

Reactive components

The important thing to realise about reactive components is that, unlike resistances, pure reactances do not dissipate power, i.e. there is no heating effect. I say 'pure reactances' since any parasitic resistive element will dissipate power and it is difficult to get a pure reactance.

Reactive components achieve this remarkable feat by reason of the current flowing out of phase with the voltage which causes it to flow. When the current is at a maximum the voltage is zero and when the voltage is at a maximum the current is zero. The net result is zero dissipation. At higher frequencies any conductor will lose power by radiating (a fact made use of in radio transmitters) but this is not a heating effect.

The difficulty with which current can flow is related to the reactance of the component. In a way directly analogous to the $V = IR$ of the resistor, $V = IX$ for reactive components, where X is the reactance of the capacitor or inductor in question. Reactance, like resistance and impedance, is measured in ohms.

Capacitors

Capacitors are conducting plates placed close together. Their capacitance value depends on how large the plates can be made and how close they can be brought together without actually bringing them into contact. The more area of plate and the closer they are together then the higher the capacitance. The capacitance also depends upon the material ('dielectric') separating the plates.

Sometimes there are several layers of plates (using both sides of the plates in the middle of the sandwich) or the plates are deposited on, or sandwiched between, thin insulating films which are then wound up tightly like a clock spring.

Often the foil is made to stick out of the end of the film, where it is joined together by a sputtering of metal. This 'extended foil' construction allows more than one contact point, in fact the foil is connected to the appropriate terminal or lead along the whole of one edge, thus minimising parasitic inductance and resistance.

Variable capacitors have one set of movable plates interleaved with a static set, or else they have interleaved layers (in older types) which are squeezed closer together with an adjusting screw.

Capacitors range in value from one picofarad (pF) or so up to many microfarads (μF). The Farad is a huge unit, which is why most practical capacitance values are measured in fractions of a Farad.

Having said that, values of several Farads have been available for some time now, but these are more like batteries and indeed they are used as such, for powering low-power memory systems, clocks and the like in computers. Don't contemplate using such capacitors as you would an ordinary capacitor.

Capacitors tend to be bulky after we get to the microfarad size. It takes quite an area of plate to get a decent capacitance and there are limits to how thin you can make the plates before they become sensible resistors in their own right! One solution to this problem has been to use an electrochemical system to maintain a thin layer of oxide as a dielectric, thereby increasing capacitance tremendously in relation to physical size.

But there are limits to the performance of these 'electrolytic' capacitors. They have a restricted operating temperature range, they lose electrolyte and age, they need to be operated with one terminal always more positive than the other (they're polarized) and they have a large ESR (see below) and poor initial tolerance. But that does not prevent them from being used as reservoir capacitors in power supplies where these parameters matter little.

Capacitors store energy in the form of an electric charge stored on the plates. Another way of looking at it is that they store energy in the 'dielectric', the insulating material or vacuum existing between the plates, in the form of a strain caused by the electric field. They tend to resist changes in voltage by soaking up charge if voltage is rising and by releasing charge if the voltage is falling.

Unlike most components, there is no electrical connection at all from one side to the other, apart from tiny leakage currents which the manufacturer tries to minimize. For this reason, current cannot flow through a capacitor in the sense which we usually think of. Direct current is stopped by the insulating barrier. Alternating current flows in a capacitor simply by virtue of repeatedly charging and discharging the plates, so that current *seems* to flow through the capacitor by merely flowing in and out of the ends!

The nature of the dielectric has a profound effect on the nature of the capacitor. The ceramic materials in particular have such huge 'dielectric constants' that capacitors formed from these materials have a very high capacitance for their volume. The snag is that the temperature dependency of the material is enormous. Capacitance changes with voltage, too, due to piezo-electric alterations in the dimensions of the material.

There may be mechanical or frictional losses in the capacitor due to the need for some polarized molecules to physically rotate as the voltage across the capacitor changes. Trapped charges contribute to losses too. 'Tan δ', the ratio of the resistive to capacitive parts of the capacitor's behaviour, is a measure of the 'lossiness' of a capacitor. Power factor is another way of expressing these losses; this time, it relates to the ratio of resistance to impedance; thus, it is a cosine rather than a tangent.

ESR is the effective series resistance, usually quoted for electrolytic capacitors. Really I suppose it should be called effective series impedance. It is a measure of the combined effects of the parasitic series resistance and inductance at a given frequency.

You will see all these figures quoted in the literature. Note, they must all be expressed at a given frequency. Quoting them on their own does not mean a lot. As a figure of merit for the capacitor, the smaller they are the better. Measurement at higher frequencies implies lower losses too.

Calculate a capacitor's reactance using $X_C = 1/2\pi fC$ where f is the frequency of the imposed waveform and C is the capacitance. With increasing frequency, the reactance falls, i.e. it's easier for current to flow at higher frequencies. At d.c. (zero frequency) reactance becomes infinite, as it does also at zero capacitance. Many waveforms are a mixture of frequencies and a capacitor (or an inductor for that matter) will have a different reactance at all of these different frequencies. You can only ever calculate for spot frequencies.

Stray capacitance is a common phenomenon wherever two wires or components are close together. Distinguish between parasitic capacitance, a (usually) undesirable feature within some component, and stray capacitance, which exists due to the proximity of other components, wiring and so on. This coupling is due to electric fields which extend from one conductor to encompass another. Shielding can help in those cases where strays are important.

There are considerably more kinds of capacitor than there are resistors, each type with its own idiosyncrasies, and it's rather an art choosing between them. Choose you must, however, if your circuit's performance is not to be menaced by peculiar side-effects.

Capacitors can leak, as mentioned previously, that is they have a very large value parasitic resistance in parallel with them. The electrolytics have poor leakage characteristics, whereas the polystyrene types are particularly good.

Unfortunately, polystyrene capacitors are not generally available at values of above about 10nF and can be a little bulky. Incidentally, the coloured end of a polystyrene capacitor shows which end is connected to the outer layer of foil. Although the polystyrene capacitor is not polarized in any way, you can gain some electrostatic shielding effect by connecting the coloured end of the capacitor to the lower impedance point of the two to which the capacitor is connected. For instance, that would be the output of an op-amp rather than one of its inputs.

Figure 6.2 shows the equivalent circuit of a typical capacitor with the parallel leakage resistance and the series inductance and resistance. The series parasitics are due to the actual resistance of the thin conductive films or foils and the act of winding the foil up into a coil. The actual values of the parasitic R and L vary considerably from one type of capacitor to another.

Capacitors have a tolerance and a temperature coefficient. Neither parameter is usually as good as those for resistors. The best temperature coefficients are to be had from the ceramic NPO

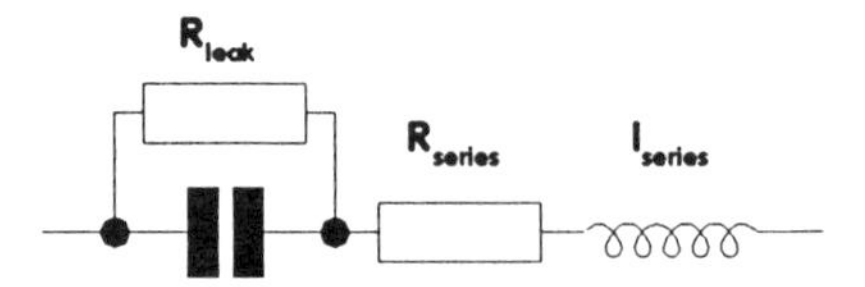

Figure 6.2: Equivalent circuit for a typical capacitor

types, the only problems here being the change in capacitance with voltage and the poor initial tolerance. The good old waxy-coated silvered mica types combine a fairly good initial tolerance with a sensible temperature coefficient, but they are expensive, bulky and not available above a restricted range of values.

Capacitors have a voltage rating which is sufficiently high for most types as to be not a problem when working at the ordinary instrumentation voltages of a few tens of volts. Look out, though, for the various electrolytic types whose voltage ratings may be smaller than most.

Table 6.1 will help you to choose between the various common types for a given application.

Capacitors are used in three ways: as filtering elements, either in power filters or signal filters, or for blocking the path of d.c.

The most common power filtering applications you are likely to come across are the very large electrolytic 'reservoir' capacitors in a linear power supply and the numbers of small 'decoupling' capacitors, usually ceramics, liberally sprinkled around any large p.c.b. which carries digital i.c.s. You could think of a capacitor in this application as 'shorting a.c. to ground', if you like.

For signal filtering, capacitors are usually associated with an inductor (typically for high frequency work as in radio receivers) or with a resistor and some active component as in audio filtering. Filters are usually incorporated as building blocks in other systems which require their services. Televisions, for example, are full of filters, filtering all manner of signals in the radio frequency, intermediate frequency (i.f.), sound and video circuits.

I suppose d.c. blocking is a kind of ultimate high-pass filter (the power filters are as low-pass as is practical). The idea of d.c. blocking is to let no d.c. through at all. Capacitors used in this

Type	Values	Tol. %	Tempco ppm/°C	Volts (d.c.)	Leakage/ Ins. res	PF, ESR, or tan δ	Uses	Advantages	Disadvantages
Aluminium electrolytic	100nF to 68,000µF	-25, +100 to ±20		10 to 450	< 3µA, mA for some	ESR up to 1.5 ohms	Reservoir, smoothing	Large values, good C/V	Leakage, ESR, polarised, limited life
Aluminium memory backup electrolytic	0.05 to 1 FARAD	-20, +80		5.5	up to 300µA	ESR up to 120 ohms @ 1kHz	Backup only (use as battery)	Large values, good C/V	Leakage, ESR
Tantalum electrolytic	100nF to 100µF	-10, +50 to ±20		6.3 to 35	up to 70µA	PF 0.2 max @ 120Hz	Smoothing, decoupling, some timing	High C/V ratio, withstands small reverse voltages	
Polyester	1nF to 4.7µF	±5 to ±20	±200	63 to 400	>10E9 ohms	PF <0.01 @ 1kHz	General	Inexpensive	
Polystyrene	10pF to 39nF	±1 or ±2	-230, -70 or ±60	63 to 630	>10E11 ohms	PF <0.001 @ 1MHz	Sample / hold	Low leakage, 1% tolerance, often good tempco	Restricted temperature range
Hi K ceramic	1nF to 22nF	-20, +80	Non-linear	63	>10E9 ohms	tan δ 6.5%	General, decoupling	Small physical sizes	Poor tempco, dC/dV
Medium K ceramic	390pF to 4n7F	±10	Non-linear	100	>10E9 ohms	tan δ 0.035	General, decoupling	Small physical sizes	Poor tempco, dC/dV
Low K ceramic	2p2F to 330pF	±2	NPO zero N150 -150 N750 -750	100	>10E10 ohms	tan δ 0.002 @ 1MHz	Temperature compensation	Small physical sizes, good stability	Available in small values only
Monolithic (multilayer) ceramic	10pF to 1µF	±10 or -20, +80	NPO +/-30 others 20 to 80%	50 to 100, some at kV	>10E11 ohms	tan δ up to 3%	Decoupling, filtering, hv smoothing	Small physical sizes, low leakage	Huge, non-linear tempco for most
Polycarbonate	100pF to 10µF	±20	±60 average	63 to 630	>10E9 ohms	PF < 0.005 @ 1kHz		Good tempco, available in larger values	Can be costly, poor tolerance
Polypropylene	1nF to 100nF	±20	-200	160 to 1500	>10E11 ohms	PF < 10E5 @ 10kHz	Pulse work	Low losses, high voltages	Can be bulky
Silvered mica	2p2F to 6n8F	±1	+35	350	>10E10 ohms	PF <0.003	Tuned circuits	Very stable	Can be costly, small values only

way are very useful for coupling amplifiers together. It often happens that there is a signal we wish to amplify whose d.c. component we are not interested in. Some amplifiers have certain requirements for 'biasing' which keeps the amplifier poised at a point where it can swing nicely either way in response to an input. The remains of this biasing appear at the output of the amplifier. Now we don't want to upset the biasing of the second amplifier so we block out the d.c. using a blocking capacitor between the output of one amplifier and the input of the next.

You can make your own very small capacitors by twisting a piece of insulated 1/0.6 wire around itself, then clipping off the top of the loop to open the circuit. Sometimes it's useful to be able to do that.

Inductors

Inductors are basically coils of copper wire. Coils with few turns might be wound self-supporting as a loosely wound but stiff coil of wire. This can be a dangerous practice if the coil then vibrates and changes its dimensions. Otherwise, inductor coils are wound on a variety of plastic cores ('formers') or a plastic bobbin with an iron core. Inductance can be adjusted in some kinds by screwing in and out an iron 'slug'. If the coils are tightly wound then the wire needs to be insulated to prevent the turns from shorting to each other. Unlike most other components, you can mostly make your own inductors if you want.

Inductors generate a magnetic field. This field can extend to include neighbouring inductors, we say that it couples to the other inductor. Sometimes this is desirable, as in the case of a transformer which is wound so the field from one coiled penetrates the other coil. Often it is undesirable. There are three things which might be done to confine the field or prevent it interfering with a neighbouring inductor.

First, separate the coils by as much distance as possible. Also check there are no ferrous metal parts which might conduct flux from one coil to another. Second, mount coils mutually at right angles to each other (you have three orientations available). Third, shield the inductor with something like mumetal or otherwise confine the flux to the core.

Practical inductance values can range from microHenrys to Henrys. Along with the desired inductance are a series resistance and an intrawinding capacitance, usually quite small. As a matter of interest, copper wire changes its resistance by about +0.4%/°C (or about +4000 ppm/°C). Also, copper changes its linear dimensions by 17×10-6/°C, resulting in a larger inductance due to expansion. Neither of these is going to have a large effect on your inductor for most applications.

Just like any other piece of wire, an inductor will get hot and eventually fuse if too much current is passed; inductors therefore have a current rating. Current handling capability is determined by the wire gauge used in the windings. This will be less than that quoted in the tables for 'free air', since the coils are tightly packed and heat cannot escape. Coils are bulky, especially if a large inductance is required along with a high current carrying capacity.

Inductors can saturate, that is they may reach a point where the domains within the magnetic material are all lined up and the material cannot contain any more magnetic flux. Where inductors carry d.c., this may be the factor which causes saturation. Inductors intended to carry d.c. often have air gaps in the magnetic circuit in order to prevent saturation. Saturation can have its uses; a 'saturable reactor' or 'transductor' is a transformer which regulates the flow of current by saturating the core from a signal winding.

A single coil used to block a.c. is called a 'choke' and is used in a similar way to the d.c. blocking capacitor. Chokes are often found in power supply wiring to prevent r.f. from getting into, or out of, a piece of equipment.

Small inductances can be obtained by the simple expedient of sliding ferrite beads (the so-called 'magic beads') over a wire. If you want to suppress r.f. pick-up this is a good idea. Gain more inductance by using a larger bead and taking several turns through the bead before leading the wire out again.

Calculate the reactance of an inductor using $X_L = 2\pi fL$ where L is the inductance and f the frequency. Increasing frequency makes it harder to pass current through the coil since the reactance increases (the opposite effect to that of a capacitor).

It has been said that electronics engineers suffer from coilophobia (an unreasoning fear of inductors!). Perhaps this is related to stray coupling problems or maybe the notion that they will have to actually wind one. Mostly, there are inductors to be had off the shelf, or at the very least, handy sizes of transformer kits to wind your own trouble-free transformers. Specialist firms will design and wind transformers for you if you really need them.

There is more to be said about choosing a power transformer in Chapter 7; the peculiarities of wiring and conductive p.c.b. tracks are treated in Chapter 8.

Winding your own inductors is a bit of a craft skill. If you must wind your own inductors, look in Chapter 14, Further Reading, for more information.

Semiconductors

Semiconductors are a real zoo. There is no recognized way of classifying them completely. A workable first approximation is to classify into discrete, digital and analogue, but there are many devices on the integrated circuit front which are both digital and analogue.

It's very tempting to launch into a detailed explanation of the inner workings of these devices. In keeping with the purpose of this book, though, I have concentrated on describing how they are used in practice. On the other hand, I have not avoided describing the underlying technology, in brief, where I feel that this will enhance your understanding.

Semiconductors are active components, that is, they are capable of amplification, of controlling power in response to a small signal, whereas passive components are mostly not. Nowadays the vast majority are made from silicon, an element derived from ordinary sand; the very first devices were made from germanium, obtained from the soot in chimneys. More recently discovered materials, such as gallium arsenide (GaAs) and its derivatives containing aluminium and phosphorus, are gaining in popularity as we learn more about their properties and how to manufacture effectively with them.

My comments on semiconductor manufacture, set out below, refer to a rather basic process. There are many other processes and variants of this process, details of which can be looked up in a number of other places if you have any particular interest. Such a process as this is used to make both discrete and integrated semiconductors.

Semiconductors are made by diffusing doping chemicals into a wafer of silicon. Diffusion takes place through a photographically produced mask which prevents the dopant from affecting areas where it is not needed. Dopant concentrations are very small, of the order of one part in a million, which means that the original semiconductor must be very pure indeed if the effects of the dopant are not to be swamped by contaminants. Confusingly enough, dopants are sometimes called 'impurities', which somehow implies that they are not wanted!

Several diffusion processes carried on one after the other result in layers of the different polarities of materials. Etching of surface features and the production of insulating areas (formed by oxidation of the silicon to silicon dioxide or glass) is also carried out at various stages. The net result is a three-dimensional pattern laid down in the very surface of the silicon wafer. Metal patterns are laid down last of all to carry the electrical connections to the edge of the chip or provide areas for bonding wires onto the unit or to connect various points within the chip to others.

Individual devices, of which there may be hundreds on a wafer, are tested and the bad ones marked. The wafer is then sawn into 'dice' using a diamond saw. Each die contains one device. The good ones are packaged; only at this point do they come to resemble their more familiar shapes (d.i.l., TO3, etc.) as actually supplied.

Doping elements are of a kind which cause excess electric charges to be forced into an otherwise neutral crystal matrix. Silicon atoms have four electrons in the outer shell and naturally form a tetrahedrally-shaped crystal. Dopant chemicals have three or five electrons in their outer shells and thus force an extra electron into the matrix or make for a missing electron (a 'hole'). These 'loose ends' are called 'carriers'.

Regions of opposite polarity *in intimate contact* form a semiconductor junction which has a barrier or 'voltage step' up which electrons can flow only with difficulty. This barrier can be swept away by forward biasing the junction (applying a more positive potential to the p-type end and a more negative potential to the n-type end) resulting in a flow of current. A junction can also be reverse biassed which increases the width of the barrier; only a tiny leakage current flows in this condition. Electric fields can also control the flow of current through the semiconductor material.

A semiconductor junction is optically active, since photons can release carriers. And like everything else we've looked at so far, it is temperature sensitive.

Semiconductors are called 'solid state' to distinguish them from valves or vacuum 'tubes'. The term solid state is actually falling into disuse, as everything nowadays is assumed to be solid state unless otherwise specified. If you want to see what a silicon chip actually looks like, look at an EPROM; the chip is directly underneath the quartz window.

We owe many of the advances in electronics today to that stroke of genius which said that the flow of current could be controlled by sweeping the charges out of a solid crystal lattice.

Diodes and diode derivatives

Diodes have two regions diffused into them. Due to the junction, current only flows easily in one direction. The two terminals of a diode are called the anode and the cathode and conventional current flows most easily from anode to cathode (in the direction of the arrow in the circuit symbol).

There is no such thing as a perfect diode, although we can come pretty close using diode/op-amp circuits. In fact forward voltage, of a small order, is needed to drive appreciable current through a diode. This voltage is often quoted as being about 0.6 or 0.7 volts for silicon and about 0.4V for germanium. This forward voltage will fall with temperature (for the same forward current) and a diode is in fact quite a good temperature sensor for the likes of an electronic thermostat.

Although this implies that there is a 'knee' in the response at this order of voltage, in fact these voltages are just a rule of thumb for moderate currents in the milliamps to tens of milliamps range. The response is in fact a smooth curve represented by a fairly simple equation. At higher currents some diodes have forward voltages of a volt or more. At smaller forward voltages currents still flow, but they are of the order of microamps or tens of microamps.

This simple one-way action, imperfect as it is, lends itself to all manner of uses. Without going into any of them in detail, logic gates, synchronous demodulators, full-wave rectifiers, absolute value circuits, mixers, detectors and clamps are all practical and possible using diodes. At small forward voltages, the diode equation approximates to a square law response and mixing or modulation is possible by superimposing two voltages on the diode simultaneously. Due to the non-linearity, multiplication happens rather than summation. Neat.

For general purposes, choose diodes according to their maximum current rating and their p.i.v. (peak inverse voltage) rating. Diodes are not expensive and you can usually afford to over-rate them by a margin. Don't do what one maker of oscilloscopes did, rating your rectifiers so close to the bone that they fail and poor old Paul has to change them all out in service. That same manufacturer, incidentally, soldered these rectifiers directly onto the power transformer; they reminded me of spiders clinging to a cliff. I love spiders, but it's a disgraceful way to mount components.

'Rectifier' is an archaic term for diode, from the days when diodes were used solely to rectify the a.c. source voltage in a d.c. power supply. It is still used but is really reserved for those diodes with an actual power handling function (as opposed to signal diodes).

Diodes fail either from excessive reverse voltage or from excessive current. The actual failure mechanism is overheating, oddly enough, rather than any other exotic effect. Over a certain reverse voltage, the diode 'breaks down'; in this state, a much higher current than the reverse current could flow. Of itself, this is not really a problem, but $P = IV$ applies. A diode carrying 10A at a forward voltage of one volt will dissipate 10W. A diode which has broken down somewhat above the p.i.v. may be carrying 100mA at 500V, so it's dissipating 50W. That's nature.

Speed is not usually a consideration except where switching supplies or logic gates are needed; use avalanche-type diodes for this. Lower forward voltage drops which minimize power losses are available by using Schottky diodes or even germanium, still available from specialist suppliers. Special low leakage current diodes are available if you're building a high impedance circuit that needs these things.

A bridge rectifier is just four diodes on one package. Make your own from four separate diodes if you want, but you'll probably find that the pre-packaged version is cheaper as well as tighter on board space, certainly for the commoner ratings.

Diodes are supposed to be noisy, but use some common sense; capacitors around a bridge rectifier in a power supply are going to do nothing for you at all.

A zener diode is just a more heavily doped diode where reverse breakdown is a feature, not a failure. A zener is intended to break down and will do so at a well defined, fairly low reverse voltage. It is deliberately used in reverse. Zeners are used for voltage regulation. Reverse voltage ratings range from about two volts to more than fifty volts. In the forward direction, a zener behaves much like an ordinary diode.

The reverse voltage of a zener rises slightly with increase in reverse current. If we divide the increase in voltage by the increase in current we get a resistance - the 'dynamic resistance' or 'slope resistance' of the diode. If the current through the zener is not well regulated then we'll get tiny shifts in voltage as the current changes. For sensitive applications a zener's reverse current needs to be kept within sensible limits.

Similar (at least in terms of behaviour) to zeners are the band-gap voltage references. Usually their voltages are of the order of one to one-and-a-half volts. These are superior to zeners in that their dynamic resistance is very low indeed and they are usually very well compensated for temperature; thirty parts per million per °C is a typical temperature coefficient for a band-gap reference whereas the temperature coefficient for a zener might be several millivolts per °C, three orders of magnitude poorer.

Band-gaps are often fabricated into an i.c. to form an on-chip voltage reference.

In days of yore there were no zeners; the breakdown characteristics of a neon would be used instead. Neons tend to break down at higher voltages than most zeners and unlike zeners they do not have a preferred forward or reverse direction. In the days of valves, a neon was a pretty handy voltage regulator.

Fast clamping diodes are available for removing transients from supply lines. They're similar to zener diodes and use the same circuit symbol, but they're much faster than zeners and can absorb the energy of the transient better.

Silicon-controlled rectifiers (SCRs) are rectifiers which can be switched on demand. As well as the anode and cathode of an ordinary rectifier, there is a gate terminal. The SCR blocks the flow of current in the reverse direction, just like an ordinary diode. It also blocks in the forward direction, until a gate voltage, positive with respect to the cathode, is applied to the gate terminal. Once triggered, the SCR conducts until it is forced off once more by the forward current falling below a threshold value (i.e. it 'latches on'). This behaviour is especially useful in a.c. circuits where the current will naturally fall to zero once every half cycle.

To switch on, the gate voltage needs to be typically about one to three volts and the current drawn from the gate circuit might be a few mA or tens of mA. SCRs of higher ratings need more powerful gate pulses; consult the data sheets for the types you have in mind. If the voltage across a thyristor rises very quickly, this can, of itself, trigger the thyristor into conduction, the so-called dV/dt effect.

Another name for the SCR is the 'thyristor'.

A triac is rather like a silicon-controlled rectifier except it will conduct both ways. The terminals are called MT1 (main terminal one), MT2 and the gate. The gate voltage is applied with respect to the MT1 terminal and can be in any sense, although the triac is most sensitive in certain directions. Typical gate voltages and currents are similar to those of thyristors.

Diacs are bi-directional trigger devices which begin to conduct above a certain voltage in either direction. About their only application is in pulse generator circuits, particularly for triggering thyristors and triacs.

Remember we mentioned the junction barrier of a diode? This barrier has a capacitance. As the reverse voltage increases so the barrier gets wider and so the capacitance gets less. The effect is tiny, but useful for r.f. circuits. In particular, 'varactor diodes' are used for tuning of r.f. front ends and for demodulating f.m., where the varactor tuning voltage required to keep the local oscillator in tune with the incoming modulation is the desired output.

Bipolar transistors

The word transistor comes from 'transfer resistor'. Bipolar transistors are current amplifiers; inject a small current into the base and a larger one is permitted to flow in the collector. I say 'permitted' since, if you have not made the necessary arrangements in the collector circuit then current will not flow. By 'collector circuit' we mean those components connected to the collector.

Kirchoff's current law comes in handy here. In the bipolar transistor, it is as though two rivers meet. The tiny tributary of current from the base meets the larger flow from the collector and joins it to come out of emitter. The emitter current equals the base current plus the collector current. A transistor is not supernatural and currents do not disappear or appear from nowhere.

This rather simple idea of the workings of the transistor will serve you well for 90% of the problems you are likely to come across in a practical sense. It is the complete answer to circuits for level shifting and d.c. amplification. Additionally, it can serve as a model for the d.c. biassing of amplifiers intended to amplify a.c.

Things will inevitably get more complicated when we start to think about the a.c. side of things, such as pulse rise times and so forth.

The word 'bipolar' means that both polarities of carrier are responsible for conduction in this device. In fact when we refer to a bipolar transistor the 'bipolar' is often dropped - 'transistor' is taken to mean good old bipolar unless otherwise stated. Bipolar transistors are three-layer sandwiches of semiconductor material and come in two flavours: n.p.n., where a layer of p-type material is sandwiched between two of n-type material; and p.n.p., which is just the reverse. Bipolar transistors therefore have two junctions.

Due to slight differences in the nature of the semiconductor materials themselves, p.n.p. transistors are not precise mirror images of their n.p.n. cousins.

The three regions are the collector, base and emitter regions, so called since the emitter emits and the collector collects electrons (for an n.p.n. transistor). The base presumably gets its name from the comparatively large piece of silicon which formed the 'base' of the transistor in the old days and into which the other regions were diffused. The base is the middle layer, the filling in the sandwich.

Remember what we said about the step, up which electrons cannot travel, in the diode. It's the same in the transistor; the collector-base junction is reverse biassed. However, when we allow current to flow in the base-emitter junction by forward biasing it, it sweeps the step away and current flows in the collector through the base to the emitter, even though the collector-base junction is reverse biassed.

The pins to which these regions are connected vary from package to package; the safest way to find out which is which is to look up in the manufacturer's or suppliers tables. Pinouts do not usually vary from manufacturer to manufacturer but will vary from type to type. A quick check with a resistance meter (use the 'diode' setting on a DMM) will at least confirm which of the connections is the base; as far as resistance measurements go, a bipolar transistor looks just like two diodes connected together at the base.

Maximum current and voltage ratings are the first things most people want to know about a transistor, followed by gain, cut-off frequency and power handling capability.

By maximum current we usually mean collector current. Any properly designed circuit will have resistances or loads which limit the collector current. Note, a transistor can actually pass far more than the maximum rated collector current; the current limits are not inherent to the transistor itself and currents must be limited by the other components in the circuit!

Gain is a very variable feature of a transistor. Gain varies between individual transistors of the same type and varies with operating conditions and temperature. Circuits in which a precise gain is important use other components to limit the overall gain of the circuit to the point where variations in the transistor's gain are immaterial. For many small signal transistors, the maximum gain occurs at about 1mA of collector current.

The collector current, otherwise known as h_{FE} bias, at which the maximum gain occurs is not often quoted in tables, but can be derived from the full data sheets.

Cut-off frequency is defined as that frequency at which the current transfer ratio of the transistor drops to 0.71 of its d.c.

value. This will vary according to the configuration of the circuit in which the transistor is being used, the common base configuration having a higher cut-off frequency than the common emitter configuration.

Power transistors are sluggish things compared with small signal devices, with cut-off frequencies in the tens rather than hundreds of MHz.

On the subject of noise, you will need to look at the full data sheets to find that current and input resistance at which the transistor contributes the least noise. Transistors of the p.n.p. persuasion are less noisy than n.p.n., due to the different way in which holes and electrons flow in the base region. But before worrying about noise too much, decide whether or not you actually need a low noise circuit. It's a thankless task minimizing the noise in a circuit when the resulting noise would not have been noticeable anyway.

Generally, circuits made from discrete transistors tend to be less noisy than integrated circuits, due to compromises in i.c. manufacture.

The most useful voltage rating is the collector-emitter voltage V_{CEOmax}. This refers to the maximum tolerable voltage stress between the transistor's emitter and collector, in the normal operating direction of the transistor and with the base open circuit.

Other parameters state the maximum collector-base voltage V_{CBOmax} (usually the same as the collector-emitter voltage) and the maximum emitter-base voltage V_{EBOmax}

This emitter-base voltage is actually a reverse voltage. When you think about it, this must be so; the normal forward voltage is of the order of 0.6 or 0.7 volts and cannot be substantially raised without a vast increase in forward current. Typical values of V_{EBOmax} for small signal transistors are four to six volts or so. Normally you would only bias the transistor in the forward direction but there are some situations, pulse circuitry comes to mind, in which a reverse emitter-base voltage might be encountered. If this is the case the voltage should be limited; a diode clamp between the emitter and base leads is a good solution.

Saturation voltage V_{CEsat} refers to the voltage remaining between the collector and emitter when the transistor is turned on as far as the circuit will permit. This happens a lot in switching circuits when the base is driven past the point where the voltage due to the collector current is dropped almost entirely by the collector load resistor. Its value will, of course, vary with collector current. It is an important factor in designing power, pulse and switching circuits.

If a transistor saturates, by the way, it will be slow to come out of that condition. Hence the Schottky TTL devices, which gain their speed by not allowing the internal transistors to saturate.

With all the parameters mentioned above, design with some margin. In particular, have regard to maximum voltage stresses, current ratings and power dissipation. Your circuit may fail to work if you ignore other parameters, but it will not blow up.

Transistors are complicated affairs and I have deliberately steered clear of detailed descriptions of circuit configurations and calculations here. There are a great many source books on the go which have plenty of examples in them; some of my favourites are listed in Chapter 14.

A transistor oddity is the unijunction transistor (UJT). It's a weird one all right, with one junction, two bases, an emitter and no collector. About the only use I've ever seen for a UJT is for a neat circuit for a relaxation oscillator. You will happen across them but rarely; I include them for the sake of completeness.

There are other exotic transistor types which you will probably not come up against, unless you are working at the leading edge of technology with microwaves and the like. Much of this research is now concentrated on the field effect transistor.

Transistors used to be optically sensitive but these seem to have gone out of fashion recently. The time was, when you could scrape the paint off an ordinary transistor and bingo! it became a useful, if somewhat crude, photodetector. I remember buying my first transistor; it was an OC71 in a metal can and it cost ten-and-sixpence. I used it to control a light bulb.

Field effect transistors

Field effect transistors (FETs) can be broadly classified into junction gate FETs (JFET or JUGFET) and insulated gate FETs (IGFET). The IGFET is also known as a MOSFET for metal oxide semiconductor FET; the oxide is the insulating layer for the gate.

Current flow in a FET is controlled by an electric field opening up ('enhancing') or closing ('pinching off') a conducting channel in a block of semiconductor material. The electric field is provided by applying a voltage to the gate contact. The gate is isolated from the conducting channel by some means so that little or no current flows between gate and channel. Just as there are n.p.n. and p.n.p. bipolar transistors, so there are n-channel and p-channel FETs.

The three terminals of any FET are called the drain, source and gate; the source sources electrons (for an n-channel device), the drain removes electrons from the device and the gate is the controlling terminal.

There is a linear (very linear in some cases) relationship between gate voltage and drain current, far more so than the distinctly non-linear behaviour of a bipolar transistor. (I use the term 'linear' here in the meaning of 'a straight line'.)

In a JFET, a gate region is diffused into a block of material of the opposite doping polarity. This forms an ordinary p.n. junction between the gate and the channel. The ends of the block are connected to the source and drain terminals. The electric field is formed by the junction barrier becoming wider as the reverse voltage is increased.

All JFETs operate in 'depletion mode', that is to say they are normally conducting but the increasing field depletes the carriers and finally 'pinches off' the channel, stopping the drain current altogether. A gate-source voltage (V_{GS}) of about three to six volts is sufficient to pinch off the channel completely in most JFET devices.

An IGFET, on the other hand, has an extremely thin insulating layer of silicon dioxide between the gate and the channel. The gate is formed from a metallization layer and the electric field extends into the channel through the insulation. IGFETs operate either in depletion mode or enhancement mode, depending on the type. Enhancement mode generates more carriers as the gate voltage increases; an enhancement mode device is normally non-conducting. Enhancement mode FETs are similar to bipolar transistors, in that the sense of their voltages and currents is the same.

The gate-source junction of a JFET must be kept reverse biassed for correct operation, otherwise it simply conducts, just like a diode, with a forward voltage of 0.6 or 0.7 volts. In other words the gate voltage of a JFET must always be outside the range of voltages between the source and drain voltages (and in the correct sense, of course). For an n-channel JFET the gate voltage must be more negative than that of either the drain or source and for a p-channel JFET the gate voltage must be kept more positive than either of these.

The IGFET on the other hand, being completely insulated, has no such restriction on gate voltage, notwithstanding its maximum ratings and its useful range of gate voltages, of course.

A FET is quite symmetrical and oddly enough will often work 'upside down' (substituting source for drain) in some circuits or, to look at it another way, will allow current to flow bidirectionally, albeit there may be some difference in forward and reverse characteristics. Some FETs are deliberately made symmetrical so that the forward and reverse characteristics are the same.

The gate of a FET draws little current (picoamperes in some cases) so it presents a very high impedance to any driving circuitry. This is the main attraction of a FET really; the driver circuitry to a high power FET does not need to be so complex as that driving a large bipolar transistor, nor is there the wasted base current, quite high in a large power transistor sometimes and certainly critical for low-power circuits. However, the gate capacitance of a FET can be quite high; driving circuitry must have current capability sufficient to charge and discharge these capacitances quickly.

Integrated circuits

There is a strong element of deferred design in integrated circuits. The designer has partitioned what he hopes is a useful set of functions or behaviours and packaged them with a well defined and simple interface.

The popularity of integrated circuits says something about their usefulness and ease of use. There is very little in the way of electronics nowadays that has not got some integrated circuitry in it.

We are less often concerned with the underlying technology in an integrated circuit than its behaviour. 'How does it behave?', we ask, or 'How is it used?', we ask, and that is the way it should be.

Analogue integrated circuits

Analogue operation implies direct representation of some measured quantity by a voltage or (less often) a current.

I suppose that all electronics is analogue in a sense; some would maintain that digital electronics is just a special case of analogue operation. Apart from the reductionism inherent in that statement, it does have some merit, in that a study of analogue electronics can bring about an understanding of the actual 'electricity' side of digital electronics, rather than the simple abstractions of 'logic'.

As soon as anyone says 'analogue' the operational amplifier, or 'op-amp', probably springs to mind as the very personification of analogue electronics. In fact there is now a bewildering array of op-amps each of which has its own peculiar features and advantages. A number of different technologies are used nowadays, with CMOS and FET types becoming more and more popular. One wonders when the shake-down will come and at what point a few generic types, each having a multitude of the advantages of existing forms, will render the older op-amps finally defunct. There are certainly devices available nowadays which out-perform by far the older types, often at no greater expense, and which are pin compatible.

All op-amps behave in a very similar way at heart. They all drive their outputs up or down according to the relative state of their inputs. Typically, they all have an inverting and a non-inverting input. Notwithstanding offset errors, if the inverting input is more positive than the non-inverting input then the output goes low; and vice versa.

An important feature of an op-amp is its gain. The popular, if dated, 741 type has a gain of about 45,000 so a change in the voltage of one tenth of a millivolt between the inputs will give a change in output of 4.5 volts. You will not often need all of this gain (!) and in actual fact the gain is controlled in most circuits by using a feedback network of some kind around the op-amp. You might think it a waste of time building a nice op-amp with some phenomenal gain just to throw it all away by restricting it using feedback, but in actual fact an op-amp with poor gain would not perform at all well; a circuit like that needs the headroom provided by the excess gain in order to work properly at sensible signal frequencies.

An op-amp has a maximum operating frequency. Its gain at d.c. is naturally limited to the quoted maximum gain, but as the frequency rises so the potential gain drops. The gain-bandwidth product is a measure of how much gain the op-amp has at any given frequency. The graph of gain *versus* frequency starts to fall off at a surprisingly low frequency (a few hertz in some op-amps) but because we have deliberately restricted the gain anyway using a feedback network, the effects are not noticeable until the gain drops to within a few times that set by the network.

The other limiting factor to speed of operation of an op-amp is its 'slew rate'. For small signals and small gains, the op-amp output can faithfully track the input. For large output voltage swings, however, the maximum slew rate limits the ability of the output to move quickly enough to continue to track the input. Slew rate limitation is over and above that imposed by gain-bandwidth product. If you bear in mind that a one hertz sinusoid of one volt peak amplitude has a maximum slew rate of one volt per second, you can simply scale up from there and find the required slew rate performance. Slew rates for op-amps are typically of the order of volts per microsecond.

All integrated circuits have input currents, however small those may be. It's tempting to think that the input current of an op-amp is zero, but for bipolar op-amps this is very often far from the case. However, it's remarkable just how little offset error there is even on an ordinary op-amp if you've taken the trouble to get the impedances the same on both inputs. By making the impedances the same, the 'input bias' currents flowing in the op-amp flow through the same impedances and generate similar offset voltages, nearly cancelling each other out.

Since we're talking d.c., we could say 'resistance' rather than 'impedance'. Basically the technique consists of choosing resistance values such that the parallel combination of all those attached to the one input is the same as the parallel combination attached to the other input. Gains are set by the ratios of resistances and it is *always* possible to alter absolute values in some sense to give a good impedance match to the other input. You should include any resistors which are attached to ground or to the op-amp output (i.e. the feedback resistor).

This is not so important in some circuits and one often sees amplifiers and filter sections with the non-inverting input connected straight to zero volts. This is acceptable if you are not worried about d.c. errors or if these errors are to be corrected elsewhere in the system.

One of my favourite little spreadsheet programs works these resistance values out for me for analogue filter circuits by suggesting a combination of resistances which are of the precisely the right ratio and absolute values. These precisely calculated values are usually really awkward so it then allows me to play with preferred values in a 'what-if' scenario until I get a good approximation to what I need. A great time-saver.

These issues are all discussed at greater length in those sections describing working circuits but they deserve to be mentioned here as they have to do with compensating for a natural feature of the op-amp itself and are potentially common features of all op-amp circuitry.

One further thing about any op-amp; it will not drive its output any further than the power supply voltages. Most op-amps will get within a volt or two of the 'rails'; some (notably the 324 type) will pull down to the negative rail but will not approach the positive rail within a volt or more; some types are known as rail-to-rail types and will drive their outputs to within a few millivolts of either supply rail . . . if there is no load on the output.

Watch the absolute value of the voltages on the inputs too. I've known some op-amps become totally deranged when faced with input voltages equal to or greater than the positive supply voltage. Some inputs allow you to equal or exceed the negative supply rail which can be handy sometimes. Look at the data sheets and check it out.

We've come a long way from the days when a single op-amp occupied a whole circuit board. I was once faced with repairing one of these many years ago; it came from a ship's engine control system. I'd never seen anything quite so complicated before and I must confess that I didn't have a clue. (I wasn't helped by the fact that there was no circuit diagram and that the manufacturer was unknown.)

Comparators are really op-amps which have been optimized for use as switches. Their gains and slew rates are higher than ordinary op-amps. Sometimes they switch so quickly that they disturb their own power lines or couple interference into their own inputs and can oscillate. Decoupling the supply with a capacitor and applying a little positive feedback can help, as can putting a modest resistance in the input leads and keeping those as short as possible.

Quite a decent comparator can be made from an op-amp with no feedback network or a network offering a little positive feedback. For non-critical applications a comparator made from a spare op-amp will do just fine.

Special op-amps are available with automatic offset nulling and precise, preset gains as well as some 'low noise' types, and so on. It's sometimes hard to tell one op-amp from another; very rarely will you have a requirement for which only one type will do. Often, some device promoted as low-noise or low-offset may have figures for these parameters which are no less than those for a device of a different category. So go by the figures and choose the one that suits your purpose; never mind what the supplier calls it.

An interesting op-amp is the auto-zero or auto-nulling op-amp. These devices measure their own offset errors, storing them on capacitors and apply them back into a nulling circuit (these capacitors may be on the chip). The long-term zero drift for these can be measured in terms of microvolts per month which is a phenomenal performance. If you're into strain gauges and other transducers which give out very small voltages, then the auto-zero op-amp is for you.

Other analogue processing i.c.s which you will come across from time to time are such things as true r.m.s. convertors, logarithmic/exponential amplifiers and four-quadrant multipliers. In a less mathematical vein, there are also integrated amplifiers for audio and r.f. work and i.c. building blocks for use as subsystems in radio and television sets.

Conversion devices

Whereas an analogue system represents, say, the brightness of a scene or the sound pressure at a certain instant by a particular voltage, digital operation uses measurements made at discrete intervals of time to represent these quantities. The essential advantage of digital working is that the signal is incorruptible within wide limits. All the analogue concerns about drift, temperature coefficient, and tolerance can to a large extent be ignored.

Long-term storage, either electronically in registers or magnetically in the form of disks, impossible with analogue technology, becomes a distinct possibility. Digital systems tend to use a lot more silicon 'chippery' than the analogue equivalent, however, and the digitized quantity suffers from 'quantization noise', a function of how many bits are used to store the measurement.

The emphasis on comparison in the foregoing might seem to imply that there is always an analogue signal as the precursor to any digital processing. Not true; digital signals are more often than not generated and used on their own with nary a sniff of an analogue conversion.

It's important to realise that we are talking about groups of signals in the digital domain. A single analogue quantity might very well be represented by eight or sixteen bits of digital information. Single digital signals have their uses too, however. These are generated, stored and transmitted using 'random logic', which is actually not random at all; the term is meant to describe the lack of bus structures and such.

Analogue to digital convertors (ADCs) and digital to analogue convertors (DACs) convert between these two ways of working. An ADC 'does a measurement', if you wish. Each measurement is called a 'sample'. There are several ways to go about conversion, each with its own advantages.

Voltage to frequency conversion (V/F) converts an analogue voltage to pulses. The V/F chip itself can be obtained in a very linear form. As voltage rises so does the number of pulses output in one second. This pulse train can be usefully transmitted over long distances without corruption as the signal can be reconstituted at the receiver end, where a counter can clock up the number of pulses received. It takes a finite amount of time to do the counting, of course; conversion is far from instantaneous.

Without distinguishing all the differences between the various other methods of conversion we can classify them all together purely in terms of their behaviour, which is: an analogue signal is input to the ADC i.c. which, after some delay, outputs a digital code or number, complete. Conversion times vary from a few nanoseconds for flash convertors to some few tens of milliseconds for the integrating types. The most common types are the successive approximation devices which will convert in as many clock cycles as there are bits in the code, plus a few; $10\mu S$ is typical for an eight-bit ADC.

DACs are usually faster than ADCs, simply because the conversion method is different. Even so, there will be a 'settling time' which may amount to a microsecond or so. Usually, as soon as a new digital code is input to the DAC, it will start to convert to the new analogue output.

What could be simpler, you may well ask? But there are one or two pitfalls set for the unwary.

First, there may be a need to distinguish firmly between analogue and digital zero volt pins. The digital zero volts is a power supply, not really a reference voltage, and noise will be coupled into the analogue side of things if these pins are joined together at this point. Where possible, take a separate analogue zero volt connection right back to the power supply terminal, sharing it with any other analogue i.c.s which need it. If the power supply is a long way from the i.c., connect a pair of diodes back to back between the pins in case the two zero volt lines should become widely separated in voltage somehow.

Sometimes you don't have access to a separate zero volts and you have to make do with a local connection. We'll be taking a closer look at these issues in Chapter 8.

Second, there is the problem of aliasing. You must not try to sample an incoming analogue waveform any less often than twice the maximum significant frequency component of the waveform. By 'significant' we mean at a level which would cause a change in the output code by one least significant bit (LSB) or more. For instance, a 1kHz sine wave must be sampled at least 2,000 times a second, and preferably more.

The problem manifests itself in that the frequency components which are too large appear in the samples as false components at a difference or beat frequency. Use an analogue anti-aliasing filter to get around this problem.

DACs have similar problems in that the output voltages are in the form of steps, so a little filtering may be needed to remove those.

A comparator can be viewed as a one-bit ADC device, if you wish.

You will probably choose to use analogue techniques where the signals you have to play with are already conveniently in analogue form and where the required output is analogue in nature. Where the frequency of operation is too great for digital sampling to work, analogue techniques are mandatory.

Digital devices

There are currently two great families of digital logic, TTL and CMOS. TTL means 'transistor-transistor logic' and CMOS means 'complementary metal oxide semiconductor'. There are other families which are ancestral to these (RTL, DTL, etc.) or which have specialized uses (ECL, I^2L) but the most common are TTL, CMOS or combinations of these two. Larger i.c.s, microprocessors and others of that order of complexity often use NMOS or PMOS, which have their own peculiar advantages, fewer steps in the manufacturing process being one of them.

CMOS when it first arrived was promoted as the 'perfect logic family' and it jolly well nearly is. It's so good at so many things that I can't see it being supplanted for some time to come. TTL is popular because it was the first comprehensive logic family and gained considerable popularity, but even it has had to make a gesture in the direction of CMOS, for instance quasi-TTL which is CMOS internally (for power saving) but is disguised as TTL for the sake of the outside world.

But what makes CMOS so good?

First, it will happily work at supply voltages from as low as three volts up to 15 volts, although there is a speed penalty at the lower voltages. This means that CMOS will run on anything from a pair of AA-type batteries, through six-volt vehicle and small boat supplies, nine volt PP3s and on through car battery-type voltages. The supply voltage does not need to be kept within close limits and voltage regulators can be eliminated for some circuits. There are one or two kinds of CMOS device which need particular supply voltages, but these tend to be low-power microprocessors and their support chips, not the ordinary logic devices.

The separate, highly stable, five-volt logic supply, required by TTL, is not always needed by CMOS. For example, if you're running analogue circuitry from a ± 7.5-volt supply and you need a modest amount of logic to go with it, then CMOS devices straddling the supply are ideal. You will probably still need to decouple the supplies, to prevent switching transients reaching the analogue circuitry, but the analogue zero volt line will remain electrically quiet as the logic is not connected to it at any point.

Second, power consumption is incredibly small. Apart from charging and discharging the gate capacitances of the on-chip MOS transistors, the only currents drawn are leakage currents. Thus quite substantial CMOS logic systems, if not being clocked, can draw mere microamps of current. The current drawn from the supply rises with clock speed (as more transitions between HIGH and LOW states are made) and is, to a very close approximation, directly proportional to clock speed. This allows the building of systems whose power consumption can be controlled by varying the clock speed.

Third, noise immunity is large as compared with TTL and is related to the supply voltage. Under no load, the output swing of a CMOS device is rail to rail, with the logic switching points at one-third and two-thirds of the supply voltage. For a 15-volt supply, voltages above ten volts are guaranteed HIGH and voltages below five volts are guaranteed LOW.

Fourth, a CMOS output can both sink and source current. It is symmetrical. The design of interfacing circuitry is very often simplified by this feature. Quite a few milliamps can be sourced from or sunk into a CMOS logic device before its output voltage begins to droop or rise to the extent where the integrity of the logic levels is compromised. Current sink and source capability rises with increasing supply voltage.

A CMOS i.c. has a high input impedance, which simplifies any driver circuitry. It also means that the fan-out of CMOS is for practical purposes infinite, as a single CMOS output can drive many, many inputs.

The down side of ordinary 'B' series CMOS is that it is not so fast as some kinds of TTL.

Early CMOS suffered dreadfully from static problems, requiring all kinds of precautions to prevent zapping. I would still recommend static precautions but it's not nearly so critical as it was in the old days.

It is worth taking a brief look at what causes static damage. CMOS, like any other MOS device, has an insulating oxide layer. This is so thin that it can be punched through by a static discharge, leaving a tiny conducting track and degrading the transistor. The trouble is, the transistor may continue to perform adequately for a while, with the damage spreading, until the transistor fails at some random and probably inopportune moment. These comments apply as much to individual MOS devices as to transistors fabricated within an i.c.

Ordinary bipolar semiconductors of the 'small junction' type are also prone to static damage if handled carelessly.

TTL is the darling of the old school, who were brought up with it before CMOS was the proverbial twinkle in the eye. The '74' series type numbers originating with TTL must be familiar sight to many involved in the electronics industry. TTL has moved with the times; there are a number of different technologies now, of varying speeds and power requirements. Some of these variants are intended to compete with CMOS, particularly in terms of power consumption. On the other hand, CMOS has fought back and there are now many CMOS devices which mimic TTL devices, at least as far as the pinout is concerned. There are even '74' series devices which are unashamedly naked CMOS but which are more or less direct replacements for TTL, which is useful if you've designed a p.c.b. which you don't want to change, but which you now want to populate with CMOS, with all its advantages.

The most commonly encountered TTL is now the LS (low-power Schottky) type; it is even more common that standard TTL. It has the same speed as standard TTL with a reduced power consumption. There is also a Schottky variant with increased speed but with a power consumption penalty. Schottky transistors are not allowed to saturate since there is a limiting diode in inverse parallel with the base-emitter junction. The diode is incorporated at manufacture by the simple expedient of extending the metallization on the chip surface to connect the base and emitter regions of the transistor and was one of the first improvements made to TTL.

There are numerous other technologies, in fact there seems to have been a recent explosion of logic families, not all TTL based, bearing the '74' series numbering system. The best advice I can give is to look up their characteristics if you have special requirements which are difficult to meet from the more usual types. Chapter 14 contains the names of various catalogues and books which contain this information.

TTL requires a five-volt power supply. It also requires that this be regulated to within ± 0.2 volts, although this is not too restrictive a constraint in these days of easy-to-use monolithic voltage regulators. The guaranteed logic LOW for a TTL device is anything below 0.8 volts. A common mistake made by users of TTL is to rate the logic HIGH voltage as five volts; it isn't. Anything above two volts is a logic HIGH; three volts is very reasonable and four volts is excellent.

TTL outputs can sink current quite readily, but cannot source much current. Buffer i.c.s can sink and source rather more, but they are still not symmetrical in this respect.

I once came across a system which used a TTL square wave as a calibration source. Apart from the fact that the supply voltage wasn't guaranteed to be anywhere near five volts (± 4%) the logic HIGH itself could have been anywhere in the range two to five volts. I implore you not to pull stunts like that; call such a device a 'checker' by all means, but don't dignify it with the name 'calibration' or base any measurement or adjustment on it.

CMOS and TTL have a many, many logic functions in common. There is a full series of small gates, AND, NAND, OR, NOR in both families, although TTL seems better endowed with the larger gates with more inputs. There tend to be more 'building block' devices (phase locked loops and the like) in CMOS than in TTL; it's easier to marry logic and analogue functions on one chip using CMOS.

You will probably choose TTL in those situations where a five-volt supply is readily available and where interfacing to existing TTL-compatible circuitry is important. You will probably choose CMOS in most other situations, especially where a five-volt supply would otherwise have to be specially provided or where power consumption is critical. Where your logic requirements are small and speed is not important (the combination of two signals which need to drive a common relay, for example) look for simpler ways of implementing any logic functions, such as using diodes or open-collector transistors.

You will probably choose to use digital techniques where long-term storage (more than a few seconds at any rate) is required, since it is difficult to preserve an analogue voltage over any period of time without corruption, due to component tolerances and leakage. You might be tempted by digital techniques where there is substantial processing or arithmetic involved, where decisions need to be made or where repeatability is paramount. Rudimentary digital electronics of some kind, even if just a switching transistor, is needed in any situation where some output device has a well defined on and off state, i.e. lamp and LED driving or relay driving.

Last but not least, digital electronics in the form of microprocessor technology offers the distinct advantages of flexibility in re-programming the circuit to perform a number of different functions, of 'deferring the design' and offloading design effort from the electronics engineer and onto the software engineer, and of introducing 'intelligent' behaviour into your products, albeit at some additional cost and complexity.

Electro-mechanical components

Electro-mechanical devices are mechanical devices that control the flow of electricity. Let's be clear on this point; things like enclosures and fans, despite being designed solely for the electronics industry, are not electro-mechanical, they're just mechanical. The best example of an electro-mechanical device is a switch of any type; relays spring to mind as a further common example. I've included a note on valves here too, just for completeness, although not may people use them nowadays.

Valves are steam-age magic. Not only do they have science fiction type structures, built to human rather than microscopic scale, tantalizingly sealed inside a gleaming glass envelope, but they glow warmly and romantically in the dark too. Compare them with the unromantic blobs of plastic or uninteresting metal cylinders which are transistors, and there's no wonder that the heart swells. Even dignified with a natty heatsink, a chip or tranny really doesn't rate.

Valves are nowadays relegated to the special jobs which they alone can handle. Technically, a transistor is so much more convenient than a valve in use (power, space and weight saving, low voltage operation and greater reliability being the most important virtues) that transistors have completely replaced valves in all situations where they could reasonably be used.

One wonders how the bipolar transistor is faring and whether the FET will do to it what the bipolar transistor did to the valve. Oddly, the FET has many of the characteristics of a valve; it is a high-impedance voltage-controlled device. In fact 'fetrons', which plugged in as direct replacements for common valve types, were around for a while, I seem to remember.

I was born just too late to really become familiar with valves; in fact (for my sins) I have never bought one or used one. They were sufficiently fascinating, however, to prompt me to collect them out of defunct television sets!

Those wonderful structures ('electrodes') inside a valve control the flow of electrons. In the inner core is a heater and a cathode. Under the influence of heat, the oxide coating of the cathode gives off electrons, forming a 'space charge' around the cathode. A negative potential on the cathode helps, too. These electrons then stream towards the positively-charged anode. If the anode does have a positive potential on it as compared with the cathode, then electrons do not flow towards it, giving a diode action.

A grid placed between these two controls the intensity of the flow of electrons (i.e. the current) by repelling them. Such an arrangement is called a triode. Several grids, all electrically isolated from each other and acting relatively independently, can be used to modulate the flow of electrons. These additional electrodes give rise to such names as tetrode, pentode and heptode. Two electrically separate valves are often placed in the same glass envelope (double tetrode, triode-pentode, etc.).

If the electrodes ever come loose, then the flow of electrons can be modulated by vibration as the electrode spacing changes. Even the voice can cause modulation. This fault is called 'microphony' for obvious reasons!

One vacuum device in common use today is the cathode ray tube. I'm looking at one right this minute. It works in very much the same way as the ordinary valve, except that the streams of electrons strike chemical deposits on the inside of the face plate. These chemicals, deposited as dots or stripes, give off light of various colours, the light intensity depending upon the electron beam current.

The cathode ray tube itself is being gradually supplanted, mainly by the liquid crystal display. Anyway, back to the less exotic.

Switches

Almost the first question anyone asks of a switch is, what should the contact configuration be?

The contact configuration is the first feature of a switch which we need to think about during the circuit design process. The operating action (rotary, toggle, etc.) and the desired effect on the behaviour of the equipment will have been laid down in the specification, but the actual contact configuration will be up to the designer.

To take toggle switches first. Due to their nature, they only have two or possibly three states. Most commonly they will be available in SPST, SPDT, DPST and DPDT forms, where SP and DP refer to 'single pole' and 'double pole' and ST and DT refer to 'single throw' and 'double throw'. 'Poles' are the number of electrically separate switches which are included in the whole switch.

Each switch section or pole has a 'common' connection which is alternately connected to one or the other of the other contacts. 'Throw' refers to the number of switch positions where this common contact is in contact with another contact. Where the switch is a simple on/off type, we say it has a single throw (even though it moves in just the same way as a double throw) since there are only two contacts on each switch, the common and one other. Where there are two contacts apart from the common (one of which is connected to the common contact at each end of the switch's travel) we say that the switch is a changeover type or double throw.

The basic toggling action of a switch can be modified (during manufacture) in two ways. First, it can be biassed, that is to say, it springs back to a preferred position when released. Second, it can be equipped with an extra centred position, in which *no* connection is made between the common terminal and either of the other two.

A switch with such a 'centre off' position and biassed one way is a frequently seen configuration and quite useful. Consider, for example, a counter system. A single switch of this kind could be used to start and stop the counter and to reset it. The centre off position could disable the counter while a biassed downward switching action could reset the counter. Leaving the switch in the upper position would enable counting. This has the advantage that the switch must pass through the 'counter disabled' position after a reset and is fine if the panel space is cramped.

Sometimes the non-common contacts are known as normally closed (NC) and normally open (NO) contacts; this terminology is more appropriate where the switch is biassed, the biassed position then being considered to be normally open since this is its 'normal' state with no operator intervention. The use of the terms NC and NO is very common in relays since by far the majority of relays are naturally biassed; NC is connected to common when the relay is not energized and NO is connected when the relay is energized, i.e. the de-energized state is held to be 'normal'.

Usually we do not refer to relays as being double throw, the term 'change-over' being used instead. Apart from this, many of the comments applied to toggle switches are directly applicable to relays.

Of course any double throw switch or change-over relay can be pressed into service as a single throw type by ignoring either the NC or NO contact. Very rarely, you will come across a switch whose contacts are like those of a toggle switch but which is actuated by a rotary movement. Circuit-wise, treat these just like toggle switches.

The abbreviation 'c/o' has fallen into disfavour. I could never tell whether the supplier meant change-over (i.e. double throw) or centre off, so as far as I'm concerned it's no bad thing. Avoid using it.

When referring to rotary switches, we tend to speak more often in terms of 'ways' rather than 'throws'. The term 'ways' refers to the number of positions which the switch can occupy. There is no such thing as centre off in a rotary switch; we generate OFF-type functions by ignoring one of the ways of the switch.

Many panel mounting rotary switches are configurable by inserting the supplied washer such that the spike on the washer stops the rotation of the switch. Some can even be made to have more ways by sacrificing a pole or two; these are usually the open wafer type.

Occasionally, one comes across a requirement for a switch which is not available from the catalogue or which seems to be prohibitively expensive. A six-pole toggle switch, biased one way and with a centre off, might be an example. If you need something like that, consider using the switch to control some switching circuit or a bank of relays instead. Alternatively, re-work the specification to a rotary type.

Apart from the configuration of the switching action, switches are usefully characterized by their mode of operation and by the current handling/voltage breaking capability of their contact sets.

Any switching mechanism, be it relay or switch proper, is characterized by its switching capacity. Usually the capacity is referred to by means of a voltage and a current. Look at it this way; the voltage is the maximum permitted open circuit voltage which would appear between the contacts were they open; the current rating is the maximum permitted current which could flow through the contacts were they closed (determined by the power requirements of the circuit being switched). Sometimes a maximum load wattage is quoted too.

No semiconductor switch can match the tremendously low on resistance of a good pair of mechanical contacts, but relays are very slow in comparison.

Snubbing networks can be used across contacts to minimize arcing. The commonest such network is a 100Ω resistor in series with a 100nF capacitor; they are often used across the main terminals of triacs too.

On one occasion I was called out to look at some relays. These poor things were trying to drive three-phase electric motors. They had been equipped with magnets between the contact sets. In theory, these magnets would divert the arc which formed, quenching it. An arc is after all a current, albeit not flowing in a wire, and as such is subject to a force when in a magnetic field. In fact this so-called 'blast magnet' can work in many cases.

However, the magnetic material itself was conductive in this case. A self-sustaining arc formed between phases when switching the motor off, going via the magnet itself, burning the contacts and splitting the magnets. When this happened, a wonderful green glow was accompanied by a lovely frying noise. After this treatment the insides of the relays had to be seen to be believed. In the event, we fitted some contactors in a separate small box and used them as amplifiers, the smaller relays controlling the contactors and the contactors controlling the motors. End of problem.

Choose the smallest relay you can which will do the job without compromising on the contact rating. Otherwise you waste space and power.

As far as relays are concerned, the larger types mount directly onto panels, mount on rails attached to the rear of enclosures (DIN rails) or plug into 'valve bases' (so called since they were the preferred valve socket) which themselves may be panel mounted or rail mounted. Smaller relays mount onto p.c.b.s by soldering in directly or by plugging into a d.i.l. socket.

This section would not be complete without a discussion of relay coils and how to drive them properly. Relays are electrically operated; a current flowing in a coil generates a magnetic field which pulls the contacts across. The exact mechanism varies from type to type. One of the advantages of using a relay is that the contacts are almost perfectly insulated from the coil circuit.

Relay coils are wound with many turns of fine wire. This is done to get as much magnetic force as possible for a given current. A current flowing twice round a loop is worth twice as much, magnetically, as that same current flowing once around. Relay designers reduce operating current to as low a value as possible by making the wire as fine as they can without reducing the current to a point where the relay will not pull in at all.

Many turns means plenty of inductance. Remember, an inductance stores energy. When you turn an inductor's current off suddenly, you get a voltage spike as the current tries to carry on flowing through the inductance. To put it another way, the magnetic field collapses as the driving current is removed; then as the lines of magnetic force cut the windings, so the collapsing field induces a voltage in the coil.

Such a spike can cause sparks, can damage driver circuitry and can radiate in the r.f. or couple to other circuits, which might get upset. Luckily, the cure is simple, at least for d.c.-driven coils; a single diode. This diode is called a catch diode, flywheel diode or freewheeling diode and is placed across the coil 'the wrong way round', i.e. so that it is not conducting when the coil is normally energized. The current which flows as the field collapses flows through the diode instead of creating a massive voltage spike in its attempts to dissipate the stored energy.

Even small relays can give a kick of fifty volts or so. Electric motors behave in the same kinds of ways and diodes around are in order there too. Some of the smaller relays incorporate such a diode and in these cases you need to be careful about which way round you apply the drive.

I remember once having to demonstrate the voltage spike to a superior of mine. I was using a solenoid valve rather than a relay, but as far as the inductance of the coil is concerned, they are one and the same thing. A certain oil company had a shut-down system on one of their platforms. It was intended to close oil wells off in the event of an accident. It used relays as the logic elements. These relays had to switch the valves on and off and they kept going on the blink. The company were getting through relays as though they were going out of fashion, and these were the expensive sealed type.

I thought that it was possible that the contacts were eroding due to sparking. To prove my point I rigged up a valve to a 24-volt supply and a switch and an oscilloscope. We had beautiful 400-volt spikes whenever the solenoid was switched off. As a result of this I advocated using catch diodes across the solenoid valves. This suggestion went down like a lead balloon; sometimes your best efforts are in vain.

Some of the really tiny relays can be operated by a naked TTL output. At the other extreme are the whacking great power relays whose coils themselves are often energized by mains voltages. Such large relays are more often called 'contactors'. They differ in appearance to ordinary relays in that the contacts are mounted at the top of the unit and are often open to the air. When the coil is energized, these contacts are pulled vigorously down (it's quite impressive to watch) to short together a static set of contacts. Three contact sets are the norm, one for each phase of the three-phase mains supply.

Often, an auxiliary set of low power contacts is provided on a contactor. These are useful for driving indicators. It's easy then for the plant operator to see that the contactor really has operated properly, important if controlling some motor far away down in the bowels of a ship, say.

Using manufacturer's data

My emphasis in this section is not to replicate the information in the manufacturer's data books but to show you what to look for in them. The extracts from the TTL Data Book and the Linear Data Book have been reproduced by kind permission of Texas Instruments UK Ltd and National Semiconductor Corporation, respectively. They follow at the end of this chapter.

I've chosen two commonly used devices to demonstrate the effective use of data sheets; the 74LS74A dual edge-triggered flip-flop and the LM324 quad op-amp.

The 74LS74A TTL flip-flop gives me the chance to wax large upon the nature of behaviour tables. The first thing you'll notice about the table is that it contains some wee arrows as well as the usual H and L for HIGH and LOW steady states.

That's because of edge-triggering; things happen in this chip at the edges of pulses. The upward-pointing arrow signifies that on the positive-going transition of the clock pulse, data is transferred from the D input to the Q output (and the complement of D, NOT D or ~D, is transferred to ~Q output).

The Xs in the table refer to 'don't care' states. Rather than listing every possible combination of HIGHs and LOWs which could exist, we can compress the table considerably by using X to mean 'this could be HIGH or LOW, since in this instance it doesn't affect operation of the chip'.

It should be clear from the table that the normal mode of operation, using CLOCK pulses and the D input, is available when both the PRESET and CLEAR inputs are in their inactive or non-asserted HIGH state.

So much for the behaviour table, the logical side of the chip's operation. The other important aspect of the chip is its electrical side. We can classify electrical parameters broadly into power requirements, input/output constraints and timing constraints.

Under recommended operating conditions, we can easily see that the required operating voltage is five volts, ±0.25V. Beware; don't read the figures for the '54' series devices which are close by. Down at the bottom of the page, we can see a figure for supply current, which is typically 4mA but which may be as much as 8mA (at 5.25V supply voltage). I would factor the larger value into your calculations for the power supply requirements, unless reliability of the finished system is not important!

Input/output constraints are expressed as a set of currents and voltages which must drive the inputs of the i.c. or which we can expect to get from the outputs of the i.c. First, note the difference between the low-level and high-level output currents, I_{OH} and I_{OL}. We can only *source* 400μA from the output of this device, but we can *sink* up to 8mA. This is very much the case, actually, for most TTL. Conversely, the current requirements to drive an input are between 20μA and 40μA maximum and 0.4mA or 0.8mA maximum for the HIGH and LOW states respectively, the symbols being I_{IH} and I_{IL}.

These figures for current are guaranteed to allow the voltages on the output pins to be representative of good logic levels, i.e. less than 0.5V for a logic LOW and more than 2.7V for a HIGH. Looking at from the other way around, if you impress these voltages on the input pins then you will need to provide no more current than that specified. In all cases, negative currents are coming out of the i.c.

Mostly, you'll be able to connect any TTL input to any output and be safe. You really need to refer to these figures if you're contemplating driving something which isn't TTL from a TTL output, or if you're taking an input from some non-TTL source, or if you have a large number of inputs connected to a single output (a 'fan-out' problem).

Timing constraints are simple for the '74' i.c., but they provide a good introduction to these things against the day when you have to delve into the inner workings of microprocessor buses! In this case we have a clock frequency, a set of pulse widths and some 'hold' and 'set-up' times.

It's fairly easy to see that we cannot run the clock any faster than 25MHz. Furthermore, the minimum pulse widths are easy to see in the table; for instance, the clock needs to spend a minimum time of 25nS HIGH for it to register reliably. With a following wind, and if it's the first Tuesday in the month, you might get away with 20nS. But don't rely on it; design safe.

Set-up times refer to how much time needs to elapse between the data on the D input stabilizing and the positive going edge of the clock pulse. In the case of the 'LS74, there is a different time for

HIGH and LOW data, 25nS or 20nS respectively. Hold times refer to how long the data on the D input must remain stable after the positive going edge of the clock has happened; 5nS in this case.

Unless you're working at very high speeds then you will not be interested in a few nanoseconds here and there. The last parameter refers to temperature; for commercial and domestic use, zero to 70°C is adequate. Specify the '54' series for an extended temperature range if you must, remembering that some of the guaranteed parameters will now be different.

The other i.c. I've chosen to look at is the '324 quad op-amp. This is a nice device for any general-purpose analogue work and has a low power consumption as well. Let's take a closer look.

The really important things for most uses are the kinds of supplies which the op-amp will run on and how close the input and output voltages can get to either of the supply rails. In the case of the 324, the power supply voltage can be anything from 3V to 30V, which implies that it will run from the same supplies as TTL or other logic. (This figure does not appear in the tables but is mentioned in the introduction to the data sheet.) Supply current is 3mA maximum, but this does not, of course, include any current supplied to the load, which may be considerably more.

Output voltage swing is zero to (V_+ - 1.5)V, in other words for a 5V supply the output voltage can reach 3.5V, a good TTL logic HIGH. Input voltages may range from zero volts (it 'includes ground' as the saying goes) to (V_+ - 1.5)V again. This is for correct operation of the device; according to the notes at the bottom of the table, input voltages can exceed this without damage (-0.3V to +32V) but the resulting output may be haywire!

Current output capability is 5mA minimum for sinking and 10mA minimum for sourcing, which implies that LEDs can be driven. These figures are for the entire temperature range; do not confuse them with the figures quoted for an ambient temperature of 25°C.

For more precise work, you will be interested in other parameters. Op-amps are not, unfortunately, ideal; they have inputs which need to sink or source current and which inevitably differ in their characteristics. Input offset voltage is a measure of how much d.c. error the op-amp introduces into the signal. You will need to multiply the V_{OS} by the gain of the op-amp stage to see what kind of effect this has on the output.

Input bias current is that current required to operate the op-amp's inputs. For the 324 this is about a quarter of a microamp, maximum, at 25°C and twice that over the entire temperature range. Flowing through a 100k resistor, this would generate a voltage of 50mV maximum. This illustrates the importance of keeping the resistances on the inputs of an op-amp the same; this issue is worked over in greater depth in the next chapter.

Coupled with input bias currents are input offset currents, or differences in the input bias current. These amount to ±150nA over the full temperature range for the 324, or about 15mV when passing through a 100k resistor. These cannot be got rid of by having equal resistances on each input. Input offset voltages have similar effects.

Lastly, you may be interested in how much noise the op-amp introduces into the signal. This will be 'excess noise', over and above any noise contributed by impedances in the circuit. It is most often specified in $nVHz^{-0.5}$ or 'nanovolts per root hertz'.

It must be said that a great many of these effects are tiny and that for many purposes they are too small to worry about. If you think that the 324 is not too wonderful (you may have applications which must work to a higher precision that the 324 can offer) then there are plenty of good op-amps around which will do admirably. If, in the light of a close scrutiny of the data, you then decide on using a super-duper amplifier, you'll have done so for all the right reasons.

Broadly speaking, use worst-case figures at all times. Steer clear of the 'typical' column in any data sheet; you need to have a really good reason for using 'typical' parameters. Otherwise, you'll eventually build a system which doesn't go, simply because the tolerances in the component parameters have conspired against you at some time. Worse yet, systems may fail intermittently in use and give you or your firm a very poor reputation.

Other comprehensive sources of component information are the catalogues of such firms as Maplin and Electromail (RS), which contain a wealth of information apart from just listing part numbers. See Chapter 14 for more on this. If you can't get a data sheet from a manufacturer or their representatives, then buy someone else's device.

54/74 FAMILIES OF COMPATIBLE TTL CIRCUITS

PIN ASSIGNMENTS (TOP VIEWS)

AND-GATED J-K MASTER-SLAVE FLIP-FLOPS WITH PRESET AND CLEAR

72

FUNCTION TABLE

INPUTS					OUTPUTS	
PRESET	CLEAR	CLOCK	J	K	Q	$\bar{Q}$
L	H	X	X	X	H	L
H	L	X	X	X	L	H
L	L	X	X	X	H*	H*
H	H	⎍	L	L	Q_0	$\bar{Q}_0$
H	H	⎍	H	L	H	L
H	H	⎍	L	H	L	H
H	H	⎍	H	H	TOGGLE	

positive logic: J = J1·J2·J3; K1·K2·K3

See pages 6-46, 6-50, and 6-54

SN5472 (J) SN7472 (J, N) SN5472 (W)
SN54H72 (J) SN74H72 (J, N) SN54H72 (W)
SN54L72 (J) SN74L72 (J, N) SN54L72 (T)

NC—No internal connection

DUAL J-K FLIP-FLOPS WITH CLEAR

73

'73, 'H73, 'L73
FUNCTION TABLE

INPUTS				OUTPUTS	
CLEAR	CLOCK	J	K	Q	$\bar{Q}$
L	X	X	X	L	H
H	⎍	L	L	Q_0	$\bar{Q}_0$
H	⎍	H	L	H	L
H	⎍	L	H	L	H
H	⎍	H	H	TOGGLE	

'LS73A
FUNCTION TABLE

INPUTS				OUTPUTS	
CLEAR	CLOCK	J	K	Q	$\bar{Q}$
L	X	X	X	L	H
H	↓	L	L	Q_0	$\bar{Q}_0$
H	↓	H	L	H	L
H	↓	L	H	L	H
H	↓	H	H	TOGGLE	
H	H	X	X	Q_0	$\bar{Q}_0$

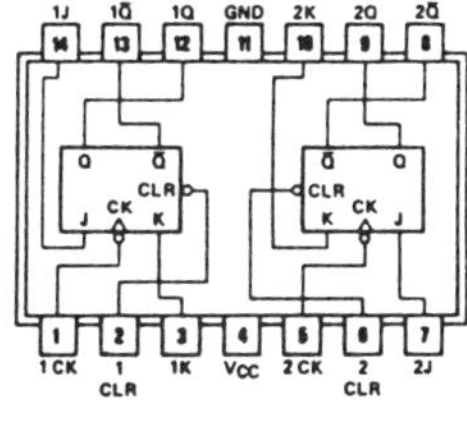

SN5473 (J, W) SN7473 (J, N)
SN54H73 (J, W) SN74H73 (J, N)
SN54L73 (J, T) SN74L73 (J, N)
SN54LS73A(J, W) SN74LS73A(J, N)

See pages 6-46, 6-50, 6-54, and 6-56

DUAL D-TYPE POSITIVE-EDGE-TRIGGERED FLIP-FLOPS WITH PRESET AND CLEAR

74

FUNCTION TABLE

INPUTS				OUTPUTS	
PRESET	CLEAR	CLOCK	D	Q	$\bar{Q}$
L	H	X	X	H	L
H	L	X	X	L	H
L	L	X	X	H*	H*
H	H	↑	H	H	L
H	H	↑	L	L	H
H	H	L	X	Q_0	$\bar{Q}_0$

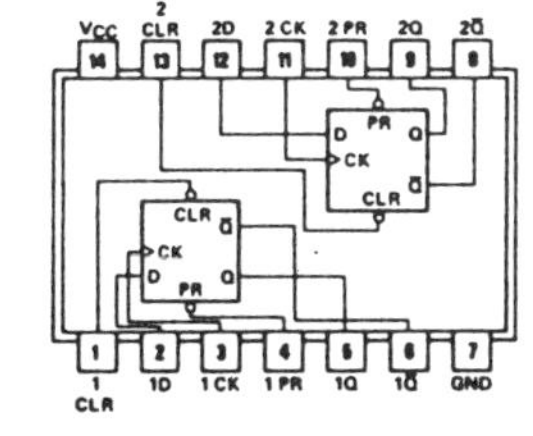

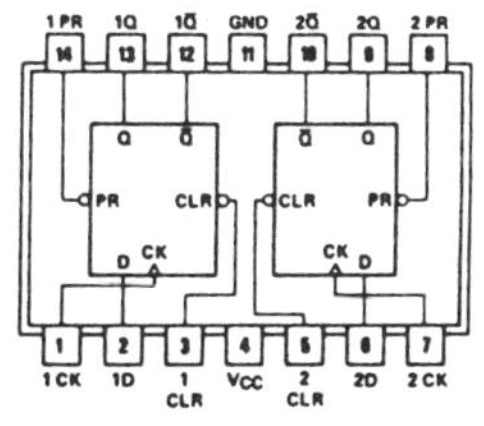

SN5474 (J) SN7474 (J, N) SN5474 (W)
SN54H74 (J) SN74H74 (J, N) SN54H74 (W)
SN54L74 (J) SN74L74 (J, N) SN54L74 (T)
SN54LS74A (J, W) SN74LS74A (J, N)
SN54S74 (J, W) SN74S74 (J, N)

See pages 6-46, 6-50, 6-54, and 6-56

See explanation of function tables on page 3-8.
*This configuration is nonstable; that is, it will not persist when preset and clear inputs return to their inactive (high) level.

SERIES 54LS/74LS FLIP-FLOPS

recommended operating conditions

	SERIES 54LS/74LS	'LS73A, 'LS107, 'LS113 MIN	NOM	MAX	'LS74A MIN	NOM	MAX	'LS76A, 'LS112 MIN	NOM	MAX	'LS78A, 'LS114 MIN	NOM	MAX	'LS109A MIN	NOM	MAX	UNIT
Supply voltage, V_{CC}	Series 54LS	4.5	5	5.5	4.5	5	5.5	4.5	5	5.5	4.5	5	5.5	4.5	5	5.5	V
	Series 74LS	4.75	5	5.25	4.75	5	5.25	4.75	5	5.25	4.75	5	5.25	4.75	5	5.25	
High-level output current, I_{OH}				−400			−400			−400			−400			−400	μA
Low-level output current, I_{OL}	Series 54LS			4			4			4			4			4	mA
	Series 74LS			8			8			8			8			8	
Clock frequency, f_{clock}		0		30	0		25	0		30	0		30	0		25	MHz
Pulse width, t_w	Clock high	20			25			20			20			25			ns
	Preset or clear low	25			25			25			25			25			
Setup time, t_{su}	High-level data	20↓			25↑			20↓			20↓			20↑			ns
	Low-level data	20↓			20↑			20↓			20↓			20↑			
Hold time, t_h		0↓			5↑			0↓			0↓			5↑			ns
Operating free-air temperature, T_A	Series 54LS	−55		125	−55		125	−55		125	−55		125	−55		125	°C
	Series 74LS	0		70	0		70	0		70	0		70	0		70	

↑↓The arrow indicates the edge of the clock pulse used for reference: ↑ for the rising edge, ↓ for the falling edge.

electrical characteristics over recommended operating free-air temperature range (unless otherwise noted)

PARAMETER		TEST CONDITIONS†	'LS73, 'LS107A, 'LS113A MIN	TYP‡	MAX	'LS74A MIN	TYP‡	MAX	'LS76, 'LS112A MIN	TYP‡	MAX	'LS78, 'LS114A MIN	TYP‡	MAX	'LS109A MIN	TYP‡	MAX	UNIT
V_{IH} High-level input voltage			2			2			2			2			2			V
V_{IL} Low-level input voltage	Series 54LS				0.7			0.7			0.7			0.7			0.7	V
	Series 74LS				0.8			0.8			0.8			0.8			0.8	
V_{IK} Input clamp voltage		V_{CC} = MIN, I_I = −18 mA			−1.5			−1.5			−1.5			−1.5			−1.5	V
V_{OH} High-level output voltage	Series 54LS	V_{CC} = MIN, V_{IH} = 2 V,	2.5	3.4		2.5	3.4		2.5	3.4		2.5	3.4		2.5	3.4		V
	Series 74LS	V_{IL} = V_{IL} max, I_{OH} = −400 μA	2.7	3.4		2.7	3.4		2.7	3.4		2.7	3.4		2.7	3.4		
V_{OL} Low-level output voltage	Series 54LS	V_{CC} = MIN, I_{OL} = MAX		0.25	0.4		0.25	0.4		0.25	0.4		0.25	0.4		0.25	0.4	V
	Series 74LS	V_{IL} = V_{IL} max, I_{OL} = MAX		0.35	0.5		0.35	0.5		0.35	0.5		0.35	0.5		0.35	0.5	
	Series 74LS	V_{IH} = 2 V, I_{OL} = 4 mA		0.25	0.4		0.25	0.4		0.25	0.4		0.25	0.4		0.25	0.4	
I_I Input current at maximum input voltage	D, J, K, or $\bar{R}$	V_{CC} = MAX, V_I = 7 V			0.1			0.1			0.1			0.1			0.1	mA
	Clear				0.3			0.2			0.3			0.6			0.2	
	Preset				0.3			0.2			0.3			0.3			0.2	
	Clock				0.4			0.1			0.4			0.8			0.1	
I_{IH} High-level input current	D, J, K, or $\bar{R}$	V_{CC} = MAX, V_I = 2.7 V			20			20			20			20			20	μA
	Clear				60			40			60			120			40	
	Preset				60			40			60			60			40	
	Clock				80			20			80			160			20	
I_{IL} Low-level input current	D, J, K, or $\bar{R}$	V_{CC} = MAX, V_I = 0.4 V			−0.4			−0.4			−0.4			−0.4			−0.4	mA
	Clear				−0.8			−0.8			−0.8			−1.6			−0.8	
	Preset				−0.8			−0.8			−0.8			−0.8			−0.8	
	Clock				−0.8			−0.4			−0.8			−1.6			−0.4	
I_{OS} Short-circuit output current*	Series 54LS	V_{CC} = MAX	−20		−100	−20		−100	−20		−100	−20		−100	−20		−100	mA
	Series 74LS		−20		−100	−20		−100	−20		−100	−20		−100	−20		−100	
I_{CC} Supply current (Total)		V_{CC} = MAX, See Note 1		4	6		4	8		4	6		4	6		4	8	mA

†For conditions shown as MIN or MAX, use the appropriate value specified under recommended operating conditions.
‡All typical values are at V_{CC} = 5 V, T_A = 25°C.
*Not more than one output should be shorted at a time, and duration of short circuit should not exceed one second.
NOTE 1: With all outputs open, I_{CC} is measured with the Q and $\bar{Q}$ outputs high in turn. At the time of measurement, the clock input is grounded.

Operational Amplifiers/Buffers

LM124/LM224/LM324, LM124A/LM224A/LM324A, LM2902 Low Power Quad Operational Amplifiers

General Description

The LM124 series consists of four independent, high gain, internally frequency compensated operational amplifiers which were designed specifically to operate from a single power supply over a wide range of voltages. Operation from split power supplies is also possible and the low power supply current drain is independent of the magnitude of the power supply voltage.

Application areas include transducer amplifiers, dc gain blocks and all the conventional op amp circuits which now can be more easily implemented in single power supply systems. For example, the LM124 series can be directly operated off of the standard +5 V_{DC} power supply voltage which is used in digital systems and will easily provide the required interface electronics without requiring the additional ±15 V_{DC} power supplies.

Unique Characteristics

- In the linear mode the input common-mode voltage range includes ground and the output voltage can also swing to ground, even though operated from only a single power supply voltage.
- The unity gain cross frequency is temperature compensated.
- The input bias current is also temperature compensated.

Advantages

- Eliminates need for dual supplies
- Four internally compensated op amps in a single package
- Allows directly sensing near GND and V_{OUT} also goes to GND
- Compatible with all forms of logic
- Power drain suitable for battery operation

Features

- Internally frequency compensated for unity gain
- Large dc voltage gain — 100 dB
- Wide bandwidth (unity gain) (temperature compensated) — 1 MHz
- Wide power supply range:
  - Single supply — 3 V_{DC} to 30 V_{DC}
  - or dual supplies — ±1.5 V_{DC} to ±15 V_{DC}
- Very low supply current drain (800μA) – essentially independent of supply voltage (1 mW/op amp at +5 V_{DC})
- Low input biasing current (temperature compensated) — 45 nA_{DC}
- Low input offset voltage — 2 mV_{DC}
  and offset current — 5 nA_{DC}
- Input common-mode voltage range includes ground
- Differential input voltage range equal to the power supply voltage
- Large output voltage swing — 0 V_{DC} to V^+ – 1.5 V_{DC}

Connection Diagram

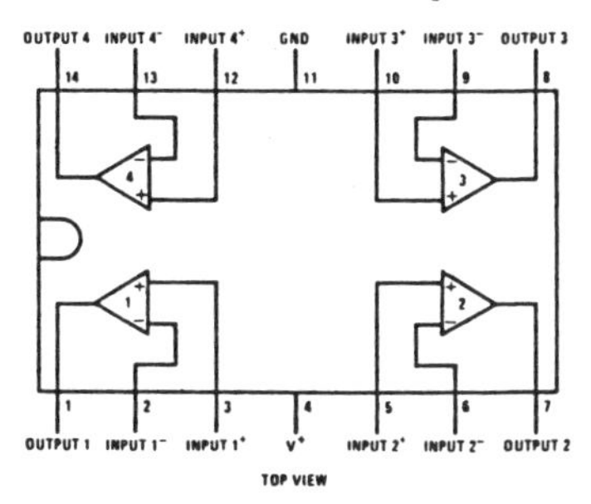

Order Number LM124J, LM124AJ, LM224J, LM224AJ, LM324J, LM324AJ or LM2902J
See NS Package J14A
Order Number LM324N, LM324AN or LM2902N
See NS Package N14A

Schematic Diagram (Each Amplifier)

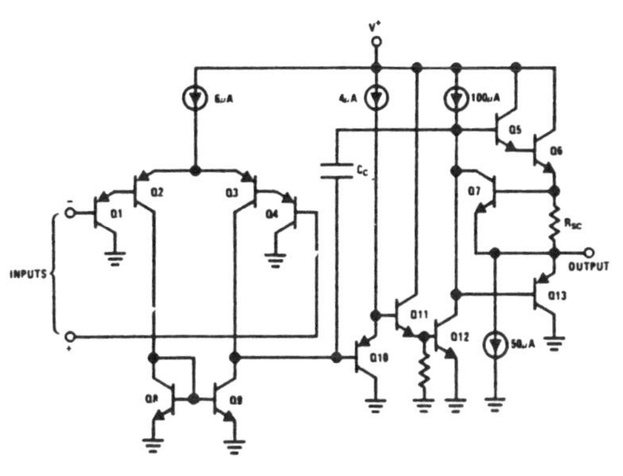

LM124/LM224/LM324, LM124A/LM224A/LM324A, LM2902

3

LM124/LM224/LM324, LM124A/LM224A/LM324A, LM2902

Absolute Maximum Ratings

	LM124/LM224/LM324 LM124A/LM224A/LM324A	LM2902
Supply Voltage, V^+	32 V_{DC} or ±16 V_{DC}	26 V_{DC} or ±13 V_{DC}
Differential Input Voltage	32 V_{DC}	26 V_{DC}
Input Voltage	−0.3 V_{DC} to +26 V_{DC}	−0.3 V_{DC} to +26 V_{DC}
Power Dissipation (Note 1)		
Molded DIP	570 mW	570 mW
Cavity DIP	900 mW	
Flat Pack	800 mW	
Output Short-Circuit to GND (One Amplifier) (Note 2) $V^+ \leq 15\ V_{DC}$ and $T_A = 25°C$	Continuous	Continuous

	LM124/LM224/LM324 LM124A/LM224A/LM324A	LM2902
Input Current ($V_{IN} < -0.3\ V_{DC}$) (Note 3)	50 mA	50 mA
Operating Temperature Range		−40°C to +85°C
LM324/LM324A	0°C to +70°C	
LM224/LM224A	−25°C to +85°C	
LM124/LM124A	−55°C to +125°C	
Storage Temperature Range	−65°C to +150°C	−65°C to +150°C
Lead Temperature (Soldering, 10 seconds)	300°C	300°C

Electrical Characteristics ($V^+ = +5.0\ V_{DC}$, Note 4)

PARAMETER	CONDITIONS	LM124A			LM224A			LM324A			LM124/LM224			LM324			LM2902			UNITS
		MIN	TYP	MAX	MIN	TYP	MAX	MIN	TYP	MAX	MIN	TYP	MAX	MIN	TYP	MAX	MIN	TYP	MAX	
Input Offset Voltage	$T_A = 25°C$, (Note 5)		1	2		1	3		2	3		±2	±5		±2	±7		±2	±7	mV_{DC}
Input Bias Current (Note 6)	$I_{IN(+)}$ or $I_{IN(-)}$, $T_A = 25°C$		20	50		40	80		45	100		45	150		45	250		45	250	nA_{DC}
Input Offset Current	$I_{IN(+)} - I_{IN(-)}$, $T_A = 25°C$		2	10		2	15		5	30		±3	±30		±5	±50		±5	±50	nA_{DC}
Input Common-Mode Voltage Range (Note 7)	$V^+ = 30\ V_{DC}$, $T_A = 25°C$	0		$V^+-1.5$	0		$V^+-1.5$	0		$V^+-1.5$	0		$V^+-1.5$	0		$V^+-1.5$	0		$V^+-1.5$	V_{DC}
Supply Current	$R_L = \infty$, $V_{CC} = 30V$, (LM2902 $V_{CC} = 26V$)		1.5	3		1.5	3		1.5	3		1.5	3		1.5	3		1.5	3	mA_{DC}
	$R_L = \infty$ On All Op Amps Over Full Temperature Range		0.7	1.2		0.7	1.2		0.7	1.2		0.7	1.2		0.7	1.2		0.7	1.2	mA_{DC}
Large Signal Voltage Gain	$V^+ = 15\ V_{DC}$ (For Large V_O Swing) $R_L \geq 2\ k\Omega$, $T_A = 25°C$	50	100		50	100		25	100		50	100		25	100			100		V/mV
Output Voltage Swing	$R_L = 2\ k\Omega$, $T_A = 25°C$ (LM2902 $R_L \geq 10\ k\Omega$)	0		$V^+-1.5$	0		$V^+-1.5$	0		$V^+-1.5$	0		$V^+-1.5$	0		$V^+-1.5$	0		$V^+-1.5$	V_{DC}
Common-Mode Rejection Ratio	DC, $T_A = 25°C$	70	85		70	85		65	85		70	85		65	70		50	70		dB
Power Supply Rejection Ratio	DC, $T_A = 25°C$	65	100		65	100		65	100		65	100		65	100		50	100		dB
Amplifier-to-Amplifier Coupling (Note 8)	f = 1 kHz to 20 kHz, $T_A = 25°C$ (Input Referred)		−120			−120			−120			−120			−120			−120		dB
Output Current Source	$V_{IN}^+ = 1\ V_{DC}$, $V_{IN}^- = 0\ V_{DC}$, $V^+ = 15\ V_{DC}$, $T_A = 25°C$	20	40		20	40		20	40		20	40		20	40		20	40		mA_{DC}
Sink	$V_{IN}^- = 1\ V_{DC}$, $V_{IN}^+ = 0\ V_{DC}$, $V^+ = 15\ V_{DC}$, $T_A = 25°C$	10	20		10	20		10	20		10	20		10	20		10	20		mA_{DC}
	$V_{IN}^- = 1\ V_{DC}$, $V_{IN}^+ = 0\ V_{DC}$, $T_A = 25°C$, $V_O = 200\ mV_{DC}$	12	50		12	50		12	50		12	50		12	50					μA_{DC}
Short Circuit to Ground	$T_A = 25°C$, (Note 2)		40	60		40	60		40	60		40	60		40	60		40	60	mA_{DC}

LM124/LM224/LM324, LM124A/LM224A/LM324A, LM2902

3

Electrical Characteristics (Continued)

PARAMETER	CONDITIONS	LM124A			LM224A			LM324A			LM124/LM224			LM324			LM2902			UNITS
		MIN	TYP	MAX	MIN	TYP	MAX	MIN	TYP	MAX	MIN	TYP	MAX	MIN	TYP	MAX	MIN	TYP	MAX	
Input Offset Voltage	(Note 5)			4			4			5			±7			±9			±10	mV_{DC}
Input Offset Voltage Drift	$R_S = 0\Omega$		7	20		7	20		7	30		7			7			7		μV/°C
Input Offset Current	$I_{IN(+)} - I_{IN(-)}$			30			30			75			±100			±150		45	±200	nA_{DC}
Input Offset Current Drift			10	200		10	200		10	300		10			10			10		pA_{DC}/°C
Input Bias Current	$I_{IN(+)}$ or $I_{IN(-)}$		40	100		40	100		40	200		40	300		40	500		40	500	nA_{DC}
Input Common-Mode Voltage Range (Note 7)	$V^+ = 30\ V_{DC}$	0		V^+-2	0		V^+-2	0		V^+-2	0		V^+-2	0		V^+-2	0		V^+-2	V_{DC}
Large Signal Voltage Gain	$V^+ = +15\ V_{DC}$ (For Large V_O Swing) $R_L \geq 2\ k\Omega$	25			25			15			25			15			15			V/mV
Output Voltage Swing																				
V_{OH}	$V^+ = +30\ V_{DC}$, $R_L = 2\ k\Omega$	26			26			26			26			26			22			V_{DC}
	$R_L \geq 10\ k\Omega$	27	28		27	28		27	28		27	28		27	28		23	24		V_{DC}
V_{OL}	$V^+ = 5\ V_{DC}$, $R_L \leq 10\ k\Omega$		5	20		5	20		5	20		5	20		5	20		5	100	mV_{DC}
Output Current																				
Source	$V_{IN}^+ = +1\ V_{DC}$, $V_{IN}^- = 0\ V_{DC}$, $V^+ = 15\ V_{DC}$	10	20		10	20		10	20		10	20		10	20		10	20		mA_{DC}
Sink	$V_{IN}^- = +1\ V_{DC}$, $V_{IN}^+ = 0\ V_{DC}$, $V^+ = 15\ V_{DC}$	10	15		5	8		5	8		5	8		5	8		5	8		mA_{DC}
Differential Input Voltage	(Note 7)			V^+			V^+			V^+			V^+			V^+			V^+	V_{DC}

Note 1: For operating at high temperatures, the LM324/LM324A, LM2902 must be derated based on a +125°C maximum junction temperature and a thermal resistance of 175°C/W which applies for the device soldered in a printed circuit board, operating in a still air ambient. The LM224/LM224A and LM124/LM124A can be derated based on a +150°C maximum junction temperature. The dissipation is the total of all four amplifiers—use external resistors, where possible, to allow the amplifier to saturate or to reduce the power which is dissipated in the integrated circuit.

Note 2: Short circuits from the output to V^+ can cause excessive heating and eventual destruction. The maximum output current is approximately 40 mA independent of the magnitude of V^+. At values of supply voltage in excess of $+15\ V_{DC}$, continuous short-circuits can exceed the power dissipation ratings and cause eventual destruction. Destructive dissipation can result from simultaneous shorts on all amplifiers.

Note 3: This input current will only exist when the voltage at any of the input leads is driven negative. It is due to the collector-base junction of the input PNP transistors becoming forward biased and thereby acting as input diode clamps. In addition to this diode action, there is also lateral NPN parasitic transistor action on the IC chip. This transistor action can cause the output voltages of the op amps to go to the V^+ voltage level (or to ground for a large overdrive) for the time duration that an input is driven negative. This is not destructive and normal output states will re-establish when the input voltage, which was negative, again returns to a value greater than $-0.3\ V_{DC}$.

Note 4: These specifications apply for $V^+ = +5\ V_{DC}$ and $-55°C \leq T_A \leq +125°C$, unless otherwise stated. With the LM224/LM224A, all temperature specifications are limited to $-25°C \leq T_A \leq +85°C$, the LM324/LM324A temperature specifications are limited to $0°C \leq T_A \leq +70°C$, and the LM2902 specifications are limited to $-40°C \leq T_A \leq +85°C$.

Note 5: $V_O \cong 1.4\ V_{DC}$, $R_S = 0\Omega$ with V^+ from $5\ V_{DC}$ to $30\ V_{DC}$; and over the full input common-mode range ($0\ V_{DC}$ to $V^+ - 1.5\ V_{DC}$).

Note 6: The direction of the input current is out of the IC due to the PNP input stage. This current is essentially constant, independent of the state of the output so no loading change exists on the input lines.

Note 7: The input common-mode voltage or either input signal voltage should not be allowed to go negative by more than 0.3V. The upper end of the common-mode voltage range is $V^+ - 1.5V$, but either or both inputs can go to $+32\ V_{DC}$ without damage ($+26\ V_{DC}$ for LM2902).

Note 8: Due to proximity of external components, insure that coupling is not originating via stray capacitance between these external parts. This typically can be detected as this type of capacitive increases at higher frequencies.

7: Design Examples

These examples are intended both to give the budding designer a taste for the design process itself and to serve as actual examples of good designs in their own right. I've gone to town on explanation, even of the most trivial detail on occasion, since I want to illustrate the richness of choices which the design environment offers.

Simplicity itself: driving an LED

A driver for a light emitting diode (LED) is indeed one of the simplest useful circuits one can imagine and yet its actual design can baffle the uninitiated. It affords an ideal opportunity to look at the decisions involved in specifying currents and voltages properly.

First of all, when entering unknown territory it makes sense to write the known design criteria down:

The problem in this case: a CMOS output must drive an LED.

The parameters: i) the LED needs 60mA forward current (it is a fibre optic communications type of LED with an appetite for power) ii) the power supply we are using is 12V.

Assumptions: The frequency at which the LED is being turned on and off is immaterial in the sense that it is sufficiently low to allow most sensible driver transistors to work well; rise and fall times are not really a problem.

A first choice needs to be made in terms of general circuit configuration. It seems obvious that the CMOS chip itself is in no way capable of either sourcing (current coming out of the pin) or sinking (conventional current flow going into the pin) 60mA for the LED. So a circuit like that shown in Figure 7.1 is probably a good first starting point, with the CMOS chip's output being buffered by an n.p.n. transistor. This is the classic common emitter interface or driver circuit which finds application in many situations.

When using any diode or indeed any active bipolar component, the starting point for any design calculations must be the voltage

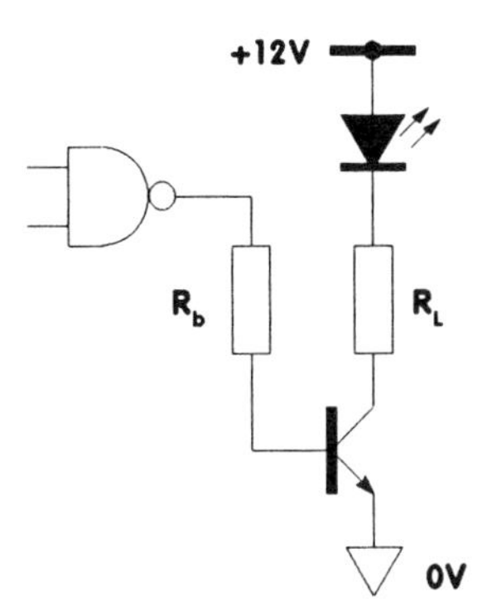

Figure 7.1: LED driver circuit using CMOS

drops inherent in the device when it is biased into a conducting condition. Taking the LED first, let us assume that it is on.

The data sheet might specify that the forward voltage is typically 1.8V when a current of 60mA is flowing. The fact that, under these conditions, there are now 10.2V total remaining across the resistor and the transistor (collector to emitter) should not baffle us. No other value of voltage is possible under these conditions.

Actually we've come to the first potential stumbling block. Note the words, 'under these conditions'. It is essential when doing any work of this kind that you can imagine the circuit in a situation where these conditions are stable. Confusion arises when we begin to say to ourselves 'but what if next . . ..' Forget it! 'But what if' might be a millisecond in the future, and we're concerned with the settled conditions prevailing right this instant. Cross bridges at the instant at which you arrive at them.

Data on the transistor might suggest that it has a saturated collector-emitter voltage of about 0.2V when turned on. This rather neatly leaves us with 10V for the resistor. Our faithful friend $V = IR$ now comes to the rescue; remember I mentioned that it always works, whatever the situation? Rearranging gives us $R = V/I$ = 10/0.06 which is 166.6667Ω.

Now we are not going to get a resistor of this value. The two nearest values in the commonly available E12 series resistors (so called because there are 12 distinct values in each decade) are 150Ω and 180Ω. The judgement part comes in deciding whether or not to drive the LED with a little less or a little more current than we originally intended.

If we work out the two possible current values for the 150Ω and 180Ω resistors (using $I = V/R$) we get 67mA and 56mA respectively. If the LED is for a display, then it will not matter if the current is a milliamp or two out. Other uses may demand a more precise value for the current. In that case, a more complicated circuit with a good constant current source might be needed. Your judgement in these matters is important; you need to bring your own experience to bear on what is essentially an exercise at the 'component choice' level. In this situation I would probably choose the 180Ω resistor and run the LED with a little less current.

What other important feature does a resistor have which we need to account for? Yes, that's it, wattage (or heat dissipation factor). $P = IV$ is the algebraic tool we need here; like our other old friend, it works in all situations. The 180Ω resistor running at 56mA dissipates 0.56W, the 150Ω resistor running at 67mA dissipates 0.67W. So we'll need a 1W resistor anyway. The 180Ω variety will run cooler; a ½W resistor incorrectly specified will get darn hot and may eventually burn out. It's resistance value will certainly change under that kind of heat stress. Specify the wattage with a certain safety margin.

An interesting point can be made here about duty cycle. Up to now, we've assumed that the LED will be switched on continuously for long periods of time. By a 'long time' we mean long enough for the load resistor to have attained a stable temperature. If, however, we are certain that (for our application) the LED will only be switched on for, say, half of the possible time, then there is an optimization which we can make.

If this is the case, then the heat dissipated will only be half that dissipated in the steady state, and we can get away with a physically smaller resistor of the same resistance value. There is one important condition which must apply to this situation; we must be able to *guarantee* that the time for which the LED is on is short compared with the time it takes the resistor to heat up, i.e. the resistor will not have time to attain what would otherwise have been its full working temperature.

'Short' probably means less than a fifth of the time that it takes the resistor to heat up. Information on thermal time constants is not always available; if it is, you can reckon that the resistor will have achieved its full working temperature after five time constants of being continuously on, and will have reached more than half of its final temperature (63%) after one time constant. Thermal time constants will usually be of the order of seconds for larger resistors or tenths of seconds for the smaller ones. So if your pulse widths were milliseconds in duration, then you would be safe enough.

If you're going to use the lower wattage resistor, you cannot tolerate odd occasions when the LED may be on for prolonged periods of time. You need to use your own judgement of reliability issues and your own assessment of the chances of the condition being violated. The safest course is to specify the more massive resistor. Perhaps there is a compelling reason for not using the more massive type, cost over 1000 units, space reasons, or whatever. Use your wits.

By the way, a bigger wattage resistor will dissipate just the same heat for the same voltage and current conditions as a less massive one. It's just that the more massive one will rid itself of that heat better. The amount of heat dissipated by a resistor depends purely on the voltage across it and current passed by it. $P = IV$ reigns supreme. We'll be exploring the heat dissipation aspects of design in more detail in Chapter 11.

That completes the collector circuit. The next big question is how we shall drive the base of the transistor to get it to switch on and pass current through from the LED. I've chosen to use a BC182L as it's a nice, inexpensive, commonly available transistor. It can handle up to 200mA and 60V (but not both at the same time, note), which is fine.

Our table of transistor parameters says the BC182L has a gain (h_{FE}) of 120 minimum. Bearing in mind that this is measured at a collector current, or 'h_{FE} bias', of only 2mA, the actual gain is likely to be less. Gain for a small signal transistor, as we have seen, tends to be at a maximum for currents in the low mA region. 100 might be a good gain figure to take. At this rate we only need 0.6mA (600μA) of base current to give us 60mA of collector current. Passing a current of 0.6mA, or more, then, into the base of the transistor will guarantee the maximum *available* current through the collector circuit.

Why maximum available current? Because the collector circuit is so arranged, as we have seen, that there is a maximum limit to the possible current flow. With the values of resistance which we have specified, and the supply voltage specified, even if we replace the transistor and LED by short circuits, we cannot possibly draw more than 12V/150Ω = 80mA. When we're trying to draw more current than can flow, the transistor is 'saturated', in other words turned on as much as is possible.

Our CMOS book of words, if we look closely at it, says that the current we can typically expect to source from a CMOS output for an ordinary 4011 type is -2.25mA. The value is negative since current is coming out of the chip. You'll find this parameter specified under 'high level output current' or similar; the symbol is I_{OH}.

This figure is quoted at a 10V supply and results in a drop of 0.5V, to 9.5V, at the output pin, due to loading of the chip's output transistors. This available current capability is certainly going to be more at the supply voltage we are using, namely 12V. Under extremes of temperature we can expect the available current to fall. However, it is guaranteed to be at least -0.9mA even at 125°C.

Now the transistor base connection will have a voltage of 0.6V or 0.7V on it when forward current is flowing. In this sense, it behaves just like a diode. I prefer to use the 0.7V figure since we're going to drive the transistor well on and also it gives us a little more margin, by using this figure we're erring on the safe side of passing a little too much base current and guaranteeing better saturation. The voltage across the base resistor R_b (for a ten volt supply) is 9.5V - 0.7V = 8.8V minimum; for a 12V supply, we can reckon on 10.8V minimum. So the base resistor needs to have a value of $R = V/I$ = 10.8V/0.6mA = 18kΩ.

To give some margin for error, it may be a good idea to source a little more than 0.6mA into the base. So if we choose the next lowest resistance value, 15kΩ, then the current we are passing is $I = V/R$ = 10.8V/15kΩ = 0.72mA, still well within the capability of the CMOS i.c. to source. The heat dissipated by this resistor is P = IV = 0.72mA × 10.8V = 7.8mW, a dissipation figure which it is hardly worth bothering with for this component.

Now how much heat is the transistor itself dissipating? Transistors have a maximum power rating, usually expressed in mW, quoted in the tables under the heading P_{TOT}. This is 300mW for the BC182L. $P = IV$ still holds true here. We have a base circuit and a collector circuit. Each of these is dissipating heat and these two dissipation figures must be added. The figure for the base side is $P = IV = 0.72\text{mA} \times 0.7\text{V} = 0.50\text{mW}$ and the figure for the collector side (a little more substantial) is $P = 56\text{mA} \times 0.2\text{V} = 11.2\text{mW}$. The final total of 11.7mW is well within the 300mW maximum.

When the transistor is off, it has a much larger voltage, 12V, across it, but the leakage current is tiny (of the order of μA) and it is unlikely that the transistor is dissipating any more than a few tens of mW.

As an aside to this question of transistor heat dissipation, consider the same transistor passing 56mA in a non-saturated condition. Ignore the base contribution, which will be the same as before, about 0.5mW. If we arrange resistance values such that only 3V is dropped by the load resistor (i.e. it only has a value of about 50Ω) then the transistor's collector-emitter voltage V_{CE} will be 12V - 3V - 2V = 7V, dissipating 392mW.

This is a parlous state of affairs for another reason; we cannot predict the gain of a transistor accurately and besides, the gain will rise with rising temperature. In this situation we have lost control of the amount of current flowing through the LED. If we have so mis-specified our load resistance value that the controlling factor becomes the transistor gain itself, then we may find that excess current flows through our expensive, shiny fibre-optic LED, killing it stone dead. You could lay bets and see whether the transistor popped first or the LED.

It is always worth looking out for excessive heat dissipation in these components. It is, after all, easily predicted. The same duty cycle criteria apply to the transistor as applied to the load resistor, except at high frequencies where the transistor becomes relatively sluggish and spends a more substantial portion of its time actually in the process of switching off. The transistor is unsaturated but still passing current during this time and excess dissipation can result.

Some rules of thumb are in order here. Typically, you can guess that the forward voltage of an LED is going to be between 1.6V and 2V. The saturated emitter-collector voltage of the transistor is usually 0.1V or 0.2V. Some tables will quote V_{CEsat} for particular currents, others won't. Gain will have to be looked up; it will be specified at a particular collector current and it will peak at a certain collector current. The important thing is

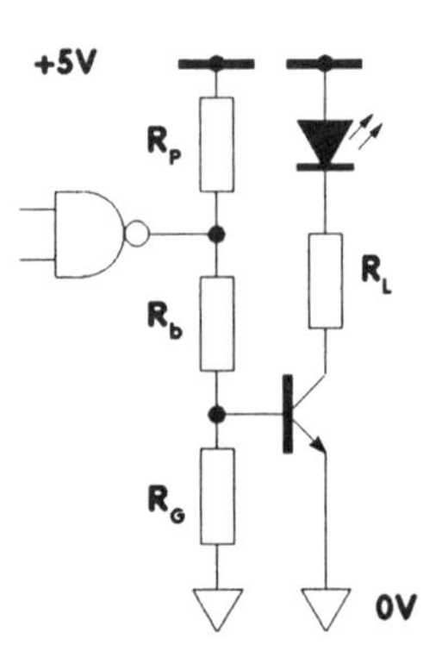

Figure 7.2: LED driver circuit using TTL

whether the circuit is guaranteed to work with the approximations we've made.

You might like to think further about using a TTL gate to drive the LED instead of a CMOS one. CMOS is simple; its outputs go all the way to the power supply rails unless we draw current and it can both sink and source current. Since the TTL output is not symmetrical, this complicates the circuit somewhat, as Figure 7.2 shows. I've assumed a 5V power supply this time, as is proper.

Before some clever person points out that we could have used a buffer of some kind, I would agree - there are some, in the Schottky TTL series or the '75' series peripheral drivers, that would just about cope with the 60mA LED current. However, we digress; the object is to show how an ordinary 74LS' chip can drive the LED.

Firstly, the TTL output cannot source more than 400μA (the usual LS TTL figure), so we need a pull-up resistor R_P to assist. When the transistor is turned on, most of the base current will in fact be sourced by this resistor, very little being contributed by the TTL output.

Secondly, the TTL output is only guaranteed to go down LOW to within 0.5V of the zero volt power rail. Now this is perilously close to turning the transistor on. Therefore, there is a further resistor, R_G, which holds the base down when the TTL output is LOW. R_P, R_G and the base resistor R_b together form a potential divider which keeps the transistor turned on until the TTL gate interferes by going LOW.

The load resistor for the LED calculates out at 50Ω now, and we can get away with a quarter-watt type. Transistor dissipation stays the same. I chose a current of 1mA to flow in the resistor chain, for no better reason than it is substantially larger then that required to flow into the base, but not too large. In that case, 280μA flows through R_G, which needs to be 2k5Ω; further calculations suggest that R_P and R_G need to have values of 1k5Ω and 2k8Ω, in order that the junction of R_P and R_b sits at about 3.5V.

With standard resistance values of 1k5Ω for R_P and 2k7Ω for R_P and R_G, we have a close approximation to this nominal arrangement, with 1.02mA flowing in the chain, 260μA flowing in R_G and the junction of R_P and R_b at about 3.47V.

When the TTL output goes LOW, the base of the transistor will be at 0.25V maximum due to the potential divider R_b/R_G, sufficient to ensure that the transistor is off. Approximately 3mA of current will flow in R_P, well within the current sink capability of the TTL i.c.

Actually we've made a meal of this example, rather flogged it to death in fact. Experienced designers, while considering all the points above in such a circuit, will dismiss some of them quickly, as their experience dictates. As your experience of design grows, you'll inevitably take similar short cuts to a sound design.

Don't forget to test the circuit on the breadboard before committing it to permanent documentation or to a soldered-up prototype. Remember, it's far simpler to debug a fragment like this at this stage than to start looking for faults when you've got fifty i.c.s on the board.

Using an operational amplifier properly

As with the previous examples, these circuits crop up again and again.

Negative feedback is the norm for all linear operation. By linear we do not mean 'a straight line' so much as 'creating no new frequency components'. Filters are a good example of linear operation. They may accentuate some frequencies and attenuate others and shift the phase of all of them, but they create no new ones. Amplifiers are (or should be) linear.

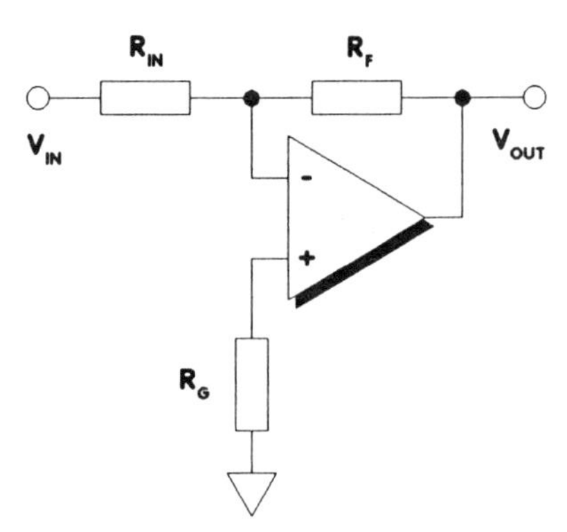

Figure 7.3: An inverting amplifier

Loosely speaking, linear has come to mean 'analogue' in a broad kind of sense as in 'linear regulator' and so on. Examples of non-linear operation are squaring, rectification, multiplication (of two signals, not by a constant which would be amplification) and modulation generally. A good way to introduce non-linear operation in any otherwise linear circuit is to drive it hard to the power supply rails, thus introducing clipping.

Take a look at Figure 7.3, which shows a simple inverting amplifier. There are one or two points which are worth explaining here.

First and foremost, what is a 'virtual earth'? The virtual earth is the inverting input of the amplifier. It's called that since it takes up a voltage very close to that of the non-inverting input, which in turn is held at zero volts reference potential, or very close to it, by the resistor R_+. (I make no apology for saying again, this zero volts connection is *not* a power supply rail.)

On its own, an op-amp will respond to voltages on its inputs by sending its output positive whenever the non-inverting input is more positive than that on the inverting input, and sending its output negative whenever the non-inverting input is more negative than the inverting input. These excursions of the output are limited by the power supply rails and the op-amp's innermost workings.

What we've just described is a comparator-type action. But when we plug in the feedback components, any tendency for the output voltage to rise is counteracted by a corresponding rise in the voltage at the inverting input, and *vice versa*.

In a normally operating linear circuit, you should see little difference in voltage between the two inputs of an op-amp. In fact, if you suspect something's wrong with some circuit you're building then it can be a good thing to look first at the inputs of the amplifier to see whether or not there are more than a few millivolts between them. This check can give you some big clues to the problem.

Let us turn to the actual feedback components themselves. Imagine that there is plus one volt (with respect to zero volts) on the input (symbol V_{IN}). The virtual earth point is held at zero volts, therefore the whole of this one volt appears across the input resistor. A current, $I_{IN} = V_{IN}/R_{IN}$, must flow in this input resistor, towards the inverting input. For the moment, let $R_F = R_{IN}$.

Apart from tiny bias currents, which we can ignore just for now, no current flows into (or out of) the op-amp inputs. The whole of I_{IN} must therefore flow in the feedback resistor R_F too (remember, currents do not evaporate into thin air). This, in turn, requires that the op-amp has driven its output down to minus one volt in order that this current can flow. It is easy to see that the circuit has a gain of minus one, and is indeed an inverting amplifier.

It's important to realise that it is the op-amp driving its output down towards the negative rail, under the influence of the voltages at its inputs, that causes these currents to flow; the current is sunk by the op-amp output and appears on the negative supply rail along with the actual supply current proper.

I've left it as an exercise for the reader to confirm that it's the same for negative voltages, when we would expect a positive output, and to imagine that R_F is twice the size of R_{IN}, in which case the circuit has a gain of minus two.

Why do we need the resistor on the non-inverting input? It's there to counteract the effects of the input bias currents. By keeping the resistances the same on both inputs, the bias currents generate the same error voltage, which thus cancels out. Its value in this instance needs to be that of R_{IN} and R_F in parallel. In general, all the resistances connected to the one input (calculated in parallel) should equal the values connected to the other (calculated in parallel).

If you're not worried about offsets, and you want to save a resistor, then by all means connect the non-inverting input directly to zero volts. But it's important that you think about it before connecting the non-inverting input in that way, rather than letting it be a default.

Say we need an inverting amplifier with a gain of -15.

First, we need to decide on the impedance of the circuit inputs. We need to choose some values of resistance which are not excessively low, loading the op-amp output, and not excessively high so that we end up with leakage, offset or noise problems. The input bias currents need to flow in the input and feedback resistors, and if the values of these are too large then the voltages at the inputs may be raised or depressed to an extent where other problems manifest themselves. There may also be a time constant due to interaction with the op-amps input capacitance, leading to an unwanted low-pass filtering action.

In general, use the smallest value resistors which do not cause large currents to flow and which do not load preceding blocks or subsystems excessively. For the 741 op-amp, for example, this will probably mean anywhere from 1k0Ω to 100kΩ. Given a free hand, 10kΩ seems to be a favourite. (In filter circuits, you may be constrained by the need to provide a resistance which gives the correct time constant when associated with a sensible capacitance value.)

If we were to choose 1k0Ω and 15kΩ for R_{IN} and R_F, this would give us the correct gain with sensible values. The maximum current which can flow with these resistance values is the supply voltage divided by the entire resistance, less than 2mA for a ±15V supply. (This worst case will only happen if the input drives up to the stops and the inverting amplifier output drives right up to the stops in the other direction.)

We must now choose a value for R_G which equals 1k0Ω and 15kΩ in parallel; my trusty Casio says 937.5Ω, an impossible value to obtain. The standard value of 910Ω will probably do nicely, as this value is a lot less critical than the other two. Even 1k0Ω would be better than a short, actually.

I'm not going to wax on about all the different possible op-amp circuits; there are excellent source books which can do far better justice to the huge number of possibilities than I can here. There are two op-amp circuits, though, which illustrate some general principles and which can be used in a number of interesting ways.

First, imagine that the circuit of Figure 7.3 now has two inputs, each with its own resistor. Each of these inputs has a gain of -1. The voltage at the output is the sum of the voltages at the inputs.

This works well for any number of inputs, although you need to watch for overloading the amplifier. You could have a giant bus with several dozens of inputs feeding to it. The virtual earth point is at a constant voltage, so the currents flowing into the virtual earth point are unaffected by currents due to neighbouring inputs, so loading of any previous stage is constant. This reduces crosstalk (a kind of interference) and is important for really precise audio work. An extra inversion can be inserted at some other point to get rid of the inversion at the summing amplifier, if needs be.

Next, consider the circuit shown in Figure 7.4; for want of a better name, I've called it a 'generalized inverse processor'. The box marked 'X' is any analogue process which we can usefully implement.

There is an interesting relationship between the action of 'X' on its own and the action of this circuit with 'X' in the feedback loop. The action of the circuit as a whole is the inverse of 'X' on its own. For example, if the feedback element were a potential divider with a tap which gave one third of the voltage, then the gain of the circuit would be three. If 'X' is an analogue multiplier set to give a squaring action, then the circuit as a whole has a square root action. If 'X' is a logarithmic amplifier circuit, the overall effect is that of exponentiation. And so on.

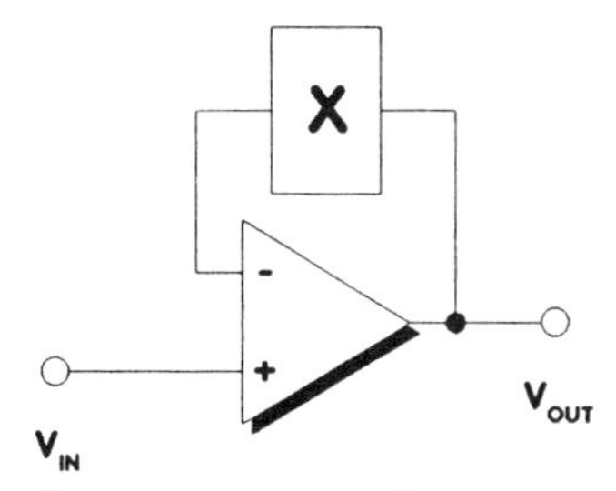

Figure 7.4: A 'generalised inverse processor'

Think about the generalized inverse processor as just another way of turning a problem on its head to make it easier to solve. The concept is based on the fact that the op-amp operates by driving its outputs to a point such that the circuit stays in balance. Sometimes this is the best way to realise some function which is difficult to make but whose inverse is easy to implement. Note that the circuit does not 'invert' in the sense of the inverting amplifier of Figure 7.3.

Double counting and the bouncy switch

This example refers to a specific problem which plagues the novice user of digital circuitry. I've seen this one so often that it merits a place in my rogue's gallery of how not to use switches and/or digital i.c.s. There are two compounding problems here but the cures are simple. This is a fine example of the use of a component as a 'safety' component.

Firstly there is the problem of the length of the leads. When the chip counts, the state changes and the power supply current will change. An i.c. with long leads may pull its own power supply down for an instant, leading to all kinds of unpredictable effects, double counting amongst them. This kind of thing is mentioned in Chapter 8. Design in a decoupling capacitor close to chip if you've got long wires going to it; by long we mean anything longer than a few inches, sometimes even less.

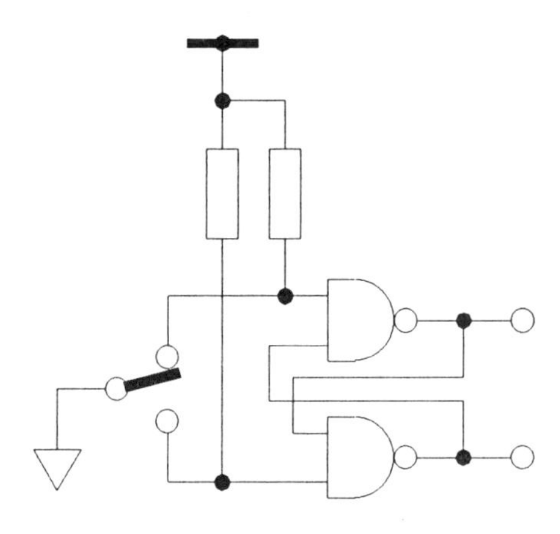

Figure 7.5: Reliable switch de-bouncing

The other problem is that of actual switch bounce itself. Ordinary mechanical switches do not close just once when they are operated; they close several times as the contacts bounce together. This results in two, three or even more pulses over a period of milliseconds. If you have changeover contacts on the offending switch then you can use the circuit of Figure 7.5 as a de-bouncing system.

This way of de-bouncing switch contacts is so reliable that it's worth specifying changeover contacts specially. Any counting or sequencing circuit attached to a switch benefits from being adequately de-bounced. As an alternative to the gates of Figure 7.5, use the SET and RESET pins of some kind of D type flip-flop.

Resistance values might be 1k0Ω to 4k7Ω for TTL. NOR gates can be pressed into service in a similar circuit, but must use pull-down rather than pull-up resistors, the common contact of the switch being connected to the positive supply.

CMOS is a better bet than TTL for the NOR circuit, since it does not need such low value pull-down resistors. Also, connecting some TTL inputs directly to the positive supply rail doesn't do them any good; LS TTL, however, doesn't mind this treatment. For pull-up or pull-down, values of up to 100kΩ are often used with CMOS.

A method for linear power supply design

Linear power supply design is one of the few major subsystems for which we can provide a design method which works for almost all cases. For most mains power supply requirements with modest currents and voltages the choices are easy to make and the parts are easily available.

This example takes each of the major components in turn and performs the calculations necessary. The idea is not so much to demonstrate the working of such a circuit as to provide some clues as to the kinds of calculations and approximations which can be used.

By 'linear', by the way, we mean 'not switching'. Switching power supplies ('switchers') are a whole other ball game.

Specification: A 9V supply supplying 1½ amps maximum, regulated to within 0.1V and powered from the standard domestic 240V mains supply. Just to be awkward we need a negative 9V

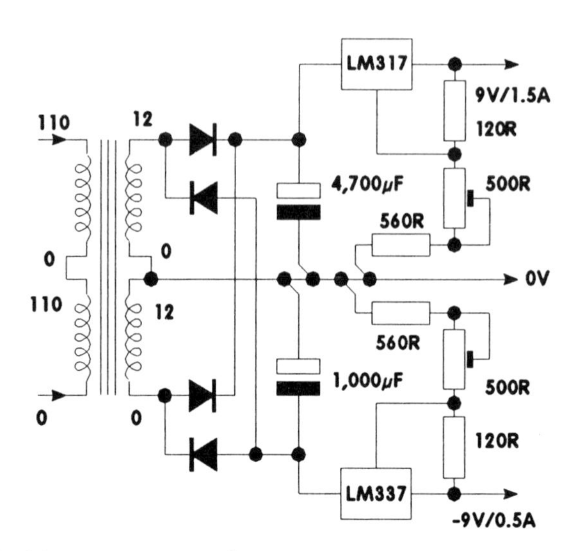

Figure 7.6: A linear power supply

supply too, with similar regulation properties, but only at half an amp.

Now 9V regulators are not easy to come by as a fixed voltage. There is an adjustable regulator which easily adjusts to 9V and can supply 1.5A, the LM317 (in either its T or K variant - we'll choose that later). Looking at the data sheets, it seems that this regulator needs up to 2.5V of headroom when supplying 1.5A if it is to regulate to within 100mV. Allowing for that, we need 11.5V minimum on the reservoir capacitor. This is a good point at which to start working.

Choosing a circuit configuration: First, there are a number of possible rectifier circuits to use, each of which has an impact on the type of transformer we can employ. To gain maximum efficiency, I am going to try one of the full wave rectifier circuits, either a full bridge rectifier or the 'biphase' arrangement. If we can get a suitable transformer with a centre tapped secondary or with two separate secondary windings which we can connect as a centre tapped winding then so much the better; we can use a bi-phase rectifier which only imposes one diode drop rather than the two diode drops of the bridge.

Better yet, I could use a full bridge circuit as two bi-phase circuits and there's then a chance that I might be able to generate my negative supply at the same time. The circuit I have sketched out in my mind is shown in Figure 7.6.

You can obtain similar circuit diagrams for other configurations of power supply circuits from various source books, or by deriving your own from this 'generic' one.

Estimating currents: Let me see what the current capability of my transformer needs to be in order to satisfy my ideal of using a bi-phase rectifier. In actual fact this happens to be 1.5A + 0.5A = 2A; for any other circuit but bi-phase I might have had to use a fiddle factor which allows for the kind of peak current loading which the particular rectifier imposes on the transformer windings.

Estimating voltages. The voltage which appears across the capacitor will be 1.4 times that of either winding (or in fact 0.7 of that across both halves in the case of a single centre tapped winding). This voltage will droop between half cycles but that is a problem to be tackled later. Our 11.5V peak translates into 8.1V r.m.s.; better account for the single diode drop at this time and specify 8.1V + 0.7V = 8.8V.

Choosing a transformer: Current × voltage is 2 × 8.8V = 17.6VA per winding = 35.2VA for both. It looks as though the nearest type is going to be one of the 50VA jobs, so we should look in that part of the catalogue. In fact there are two transformers which might fit the bill; one with 9V at 2.7A per winding and one with 12V at 2A per winding. Certainly with the 9V variety everything will run cooler and we will have some current headroom to boot. So let us provisionally choose that one and see whether or not, in the event, it is the one to go for.

Choosing a rectifier. 1.4 × 9V = 12.6V p.i.v. Bear in mid that each of the windings only conducts for half the time with a bi-phase circuit so each rectifier needs to be rated only 0.75A for the positive side and 0.25A for the negative side. The p.i.v. is very relaxed so if we specify a 1A bridge with a p.i.v. rating of 50V (the lowest available at 1A) we have some headroom for current and lots to play with as far as p.i.v. goes. In fact this p.i.v. rating will accommodate the higher voltage transformer too, so unless something really unexpected happens we have found our rectifier.

Calculating capacitance values. The capacitor voltage will ripple. The capacitor is intended as a reservoir (hence the name) which will hold up the supply between half-cycles of the mains. The voltage must not droop below 11.5 volts at any time, otherwise the regulator stands a chance of no longer regulating.

Now the capacitor voltage will peak at about 1.4 × 9V - 0.7V = 11.9V. This only gives us 0.4V of possible ripple voltage before the regulator stops regulating properly. That looks mighty close; working out the capacitance using a value of 10mS for the time between charging up every half-cycle (i.e. we are assuming 50Hz operation):

$$C = it/V$$

gives us

$$C = 1.5\text{A} \times 10\text{mS} / 0.4\text{V} = 37{,}000\mu\text{F}$$

Even given that the value of 10mS is a conservative rating, this capacitor is still a corker; a beezer, in fact, which could cost us five or ten pounds.

Let's go round the loop once more and see if the transformer with the higher voltage rating helps out at all. We now have 1.4 × 12V = 16.8V to play with. Subtracting the minimum voltage gives 16.8V - 11.5V = 5.3V and the required reservoir capacitance becomes

$$C = 1.5\text{A} \times 10\text{mS} / 5.3\text{V} = 2{,}800\mu\text{F}.$$

What a difference! We can choose a 4,700μF, 25V capacitor which will probably cost us less than two pounds and gives us plenty of headroom, both in terms of a conservative capacitance value and a relaxed voltage rating.

Using similar techniques we can work out the capacitance needed on the negative side as being 1,000μF; try it yourself and see. I've inserted these values into the diagram.

There's one loose end to tie up; the resistances for adjusting the regulator voltage. The data sheets for the regulators suggest that the resistor from the output terminal to the adjust terminal can be 120Ω in both cases. So then we need to solve the equation

$$V_{OUT} = 1.25 \times (1 + R / 120)$$

for R. This rearranges, in the manner I've described previously, to

$$R = 120 \times ((V_{OUT} / 1.25) - 1) = 120 \times 6.2 = 744\Omega$$

This resistance is made up from a fixed resistor plus half of a variable resistor (assuming that the wiper will be nicely positioned halfway along the potentiometer track when the regulator has been adjusted). If we make a further calculation based on an output voltage of 8V, we get

$$R = 120 \times ((8 / 1.25) - 1) = 120 \times 6.4 = 648\Omega$$

which is the value of the fixed resistor which would enable an adjustment to be made down to 8V. But let's give ourselves a little extra leeway; let's choose the nearest standard value less than this, 560W, giving a lower voltage adjustment of

$$V_{OUT} = 1.25 \times (1 + 560 / 120) = 7.1\text{V}$$

Doubling the difference between our fixed resistor value and the nominal total resistance value gives

$$R_{POT} = (744 - 560) \times 2 = 368\Omega$$

which is the value needed for the potentiometer. Dare I suggest that a 500Ω pot would do nicely here? The only provision being, not to wind it up to the very top, as the power supply would then not regulate properly. Bear in mind that this is a design for a fixed 9V power supply.

There are one or two components which I've left out of the diagram for clarity. If you're keen to build this power supply, remember that you will need a switch and a mains fuse. Also, there is the odd capacitor needed for stability and for greater ripple rejection, and a possible need for protection diodes; look up the data sheets for the regulators.

The power dissipation which we can expect from such regulators, and what to do about it, is covered in Chapter 9.

Other sources of circuit information

It has been said, again I can't remember where exactly, that plagiarism is the best design technique. Up to a point, I would agree with that statement. Tried and tested circuits are a boon, especially if you're in a hurry. No-one can object to you coming under the influence of their published material (that's what teaching is about, after all). On the other hand, wholesale copying of circuits, systems and subsystems, values and everything, is not an option.

One thing which is definitely worth avoiding is the 'not invented here' syndrome, the desire to be different, whatever the cost. There are hundreds of generic circuits on which no-one has a monopoly or patent. Adapt them to your needs. If someone else does it better than you at a price you can afford, buy it in. After all, very few firms make their own chips, for example.

Interpret circuit diagrams from source books carefully. There are two pitfalls for the unwary. First, you may not know what kinds of circuits the example is driven from or what it is intended to feed; look carefully at whatever you intend to hook it up to. Second, the circuit may not behave in exactly the way you'd intended; you'll need to experiment with variants.

One common problem can be that the various fragments you've collected all need different supply voltages. To summarize, you've got some work to do to persuade the bits and pieces to co-operate.

Like the circuit in Figure 7.6, circuits in source books, cookbooks or applications handbooks are just suggestions and you will have to use your calculator and your noddle to turn them into the real, working circuits that you need.

Study successful designs at every opportunity and allow them to influence your decisions. Any circuit design will be a blend of tradition and innovation; draw the line in the place that suits you best.

8: Power Supplies, Earthing and Noise

Poor power supply specification and bad wiring can be the bane of any project or product. Like the mechanical side of electronics, wiring and power supplies are often ignored in the hope that everything will come out in the wash and be all right at the end of the day. This is a pity, since an otherwise brilliant project can be rendered useless by lack of attention to wiring and power.

I've grouped power supplies, earthing and noise together in a single chapter as any problems with the one tend to compounded by the rest. (The kinds of noise I will be referring to in this chapter are not thermal or excess noise, but those unwanted signals or interference injected into a circuit by poor wiring layout.)

There are safety aspects to earthing and wiring, too, which are addressed at length.

On the importance of power supplies

Without plenty of stable power, many circuits fail to come up to expectations performance-wise. With few exceptions, even those circuits which seem to thrive on a multitude of supply voltages have some form of internal regulation to give the rest of the system the clean, stable power it needs.

One of the most basic choices we can make about power supplies is whether to use battery or mains power. Often, the choice is made for us. On the one hand, portability may be a definite requirement; on the other, the system may be a permanent installation, needing a lot of power and within easy reach of the mains plug.

Batteries are eventually going to run out. Battery lifetime is a big issue, not only from the point of view of the battery going flat in use, but also from the point of view of how many cycles of charging and discharging the rechargeable (secondary) cells will withstand.

Battery power from dry cells, in terms of energy per $, is the most expensive kind of electricity you can have. Check to see whether battery power is really needed, then if you decide that it is, choose a battery pack or system carefully. Do your users a favour by trying to minimize power consumption, lessening the expense of replacing batteries. Another tip for the battery users is to avoid, where possible, converting from one voltage to another. Design for batteries is eased by having a single supply rail wherever possible.

With any kind of power supply, battery or otherwise, check that the available current is adequate.

After a discussion of wiring, we'll return to power supplies to see how they may best be connected up.

Earth, ground, chassis, zero volts and common

It does no harm to sort out the above terms. Some of them are used rather loosely, which is quite hazardous if you get it wrong and connect something up in an unsafe manner.

Earth is the middle pin of the mains connector, which is connected to a cold water pipe or to a metal spike driven into the ground just outside your house or in the 'solum', that bit of dead space under the floor of your house which isn't quite a cellar. It is, quite literally, earthed.

Sometimes an earth connection is called 'ground'. Used strictly as a synonym for earth, the term 'ground' is OK. But don't introduce ambiguity by confusing 'ground' with 'zero volts'.

Neutral is a kind of earth but earthed back at the last electricity sub-station or transformer before your house. It actually carries the return mains current back to the power station, unlike an earth proper which carries no current unless there is a fault somewhere. I've described the consequences of ignoring this difference under use of the oscilloscope in an earlier chapter. You won't blow a fuse if you connect earth and neutral together, but you may set something on fire.

A residual current circuit breaker (r.c.c.b.) can be used to get round this problem. It works by sensing the currents flowing in the live and neutral lines. If they differ by more than a few tens of mA (30mA is typical) then the circuit is disconnected. If you connect yourself between live and earth, the usual story, an r.c.c.b. can save your life. If you have the misfortune to connect yourself between live and neutral, you have a more difficult time; an r.c.c.b. will not see the difference between you and the system being powered. Currents flowing between earth and neutral will also pop the breaker.

An earth-free zone is an area from which earths have been deliberately excluded. An earth-free zone is quite safe really since you would have difficulty connecting yourself between live and neutral. Power is brought into such a zone via an isolating transformer, such that there is a voltage difference to drive instruments and so on. An accidental single point connection draws no current apart from a leakage current due to the capacitance between windings.

In a medical setting, the isolating transformer comes into its own. Anything which is to be attached to a human being and which is mains powered has to have a medical grade transformer which allows no leakage from primary to secondary. The insulation on such a transformer is to a very high standard and the capacitance between the windings is small by virtue of a screen wound between the primary and secondary.

These precautions are needed because the patient is often well and truly earthed (by virtue of lying on a metal table covered in saline solution etc.) and because the probe of the instrument may be in intimate contact with the patient's inner workings. A current as low as $10\,\mu$A injected to the heart can cause the spasm of the heart muscle known as fibrillation.

Zero volts is, literally, a point from which all other voltages are measured. It means no more than that; one should not assume that the zero-volts point in a system is actually earthed.

Sometimes zero volts is called 'common'. Really, common is anything physically common, i.e a common zero volts, a common 24 volt rail, the common point at which two things join. Don't always assume that 'common' means 'zero volts'.

Chassis is not a term often used nowadays but it refers to the metalwork in the unit. Don't assume that 'chassis' means the same as 'earth'. Inevitably, any exposed metalwork, the chassis, on a modern instrument is earthed unless the equipment is battery powered and/or the exposed metal is very well insulated from any of the internal electronics (doubly insulated). In the older television sets, however, the chassis was at half mains voltage; the viewer was protected by a nice wooden cabinet.

An old friend of mine still bears the scars to remind him of the time he was working on such a chassis. The set was switched on; he realised very soon after picking it up that he had to put it down again as quickly as possible.

Earthing

There are some complicated rules about how to wire earthing and there are some complicated exceptions too. But there are three common practices that we can choose from and there are some wiring ideals which we can try to live up to.

First, earthing of all metal parts. This is the case for equipment powered from the mains and which is contained in a metal box. Common examples are the oscilloscope and the desktop computer. Paint, by the way, is not counted as protection; even if the metal is coated with what appears to be a nice thick insulating layer of paint, it must still be earthed.

Unfortunately, this system means that signal integrity may be compromised due to earth loops, especially with co-axial connectors like the BNC type, which usually have their outer parts connected to the external metalwork (and thus presumably earthed). Whenever we connect two pieces of mains-powered equipment together we're in peril from the earth loop.

Electromagnetic induction causes currents to flow around this loop. Where this loop is small (the equipment is powered from the same socket or is in the same room) and the systems are digital this will not matter a great deal, which is why all those desktop computers in the world today still work, even when hooked up to a printer.

Analogue systems will suffer far more; humming audio systems are typical of the earth loop problem, as currents flow around the loop through the signal return, electrostatically impressing a tiny hum voltage in the signal wire itself. In sensitive laboratory instrumentation it can be a real pest.

Figure 8.1 shows how an earth loop can be formed. The two instruments, each consisting of a power supply and other circuitry, are plugged into the mains supply and connected to each other. Current can flow from earth, through the power supply wiring, through the circuit board itself, via the signal return wire and then via a similar path in the other instrument. Note that there are other earth loops too, within the instruments themselves, which are due to the connector outers making contact with the case. Thus

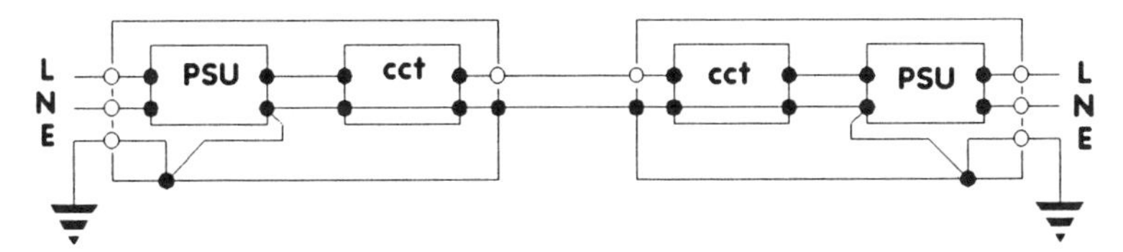

Figure 8.1: Earth loops

signal currents travel through the case of the instrument, not a desirable state of affairs.

There are several ways out of this earth loop problem.

One might isolate the instruments electrically ('galvanic isolation') from one another. This is the best possible thing to do but is only easy for digital signals running at modest speeds. The MIDI interface is the most popularly known way of isolating equipment while allowing signals to proceed; it uses optoisolators. Other options for galvanic isolation might be to use actual fibre-optic cables or isolating transformers.

You could transmit data in true differential form rather than having an 'earthy' side. This pushes the circuit complexity up a little, as there needs to be a transmitter and receiver, although these, at their simplest, may need to use no more than an op-amp and a few resistors apiece. This can be done quite simply for analogue signals; the digital equivalent is RS422, cousin to the widely known RS232.

We could break the loop within the instrument by using an isolated BNC connector. This doesn't stop the connector outer from being earthed, but it provides just one signal return path, eliminating signal currents in the enclosure of the instrument, at least.

Earthing the box doesn't necessarily mean that the internal circuit needs to be earthed too. It can float instead. (Where the circuit is earthed, it is zero volts which is connected to earth.) It is usual in this situation to provide terminals at the rear of the enclosure, on a standard ¾" spacing, one for zero volts and one for earth. These can be strapped together in the event that an earth is needed at this point.

Let me tell you about a couple of spectacular earth loop problems I witnessed.

The first was that of a modem intended to transmit data to a central computing centre. All went well until a graphics tablet was hooked into the computer too. When the graphics tablet was plugged into the mains, the modem appeared to go haywire and data could not be transmitted. A glance at the scope showed RS232 data transmission mixed with a 50Hz sinusoid of about the same amplitude. When the graphics tablet was unhooked either from the mains or from the computer all was well.

Now the modem was actually several rooms away from the computer and across a corridor and it was also plugged into the mains. It was a severe case of earth loop; the earths at the graphics tablet and at the modem being connected via a long piece of wire, probably all the way back to the distribution box. The signal grounds in the modem and graphics tablet cables completed the loop, when both were plugged in. Rather than unhooking earths, we left the operator with instructions to unplug the graphics tablet when not in use; she couldn't use them both at the same time anyway.

The second was off-shore. A colleague of mine had unhooked two ends of a zener barrier safety earth from each other, suspecting there was another unofficial connection elsewhere causing a loop. Due precautions being observed, I might add. He was using one of the old AVO 8's and could not get a sensible resistance reading to check for a short circuit. He went on to check how much current was flowing in the circuit. Imagine our surprise when the AVO popped the cut-out on the 10A scale! All of this current, as far as we could gather, was being induced in the loop. In the end I think he discovered that the two rails had been separately earthed already, at a secure but remote point, and that the two earth bars were inappropriately connected inside the control panel.

In the extreme, people have rigged oscilloscopes with no earth to get rid of a loop, either by disconnecting it in the plug or (better) by using a brightly coloured socket outlet with the earth left unmade. This is not a practice I would recommend, since the braid in the probe lead is hardly of sufficient size to take a fault current reliably to earth.

Problems are compounded when the scope is no longer connected to the intended system under test; this leaves us with a floating metal case surrounding mains powered circuitry. Systems which rely on the integrity of the ordinary power transformer insulation alone ('basic insulation') for protection must have exposed metal parts earthed.

The second common practice we can use is 'double insulation'. If the circuitry itself is enclosed in an insulating box, then we have another layer of insulation ('supplementary insulation') and can claim double insulation protection. Exposed metal parts which lie completely outside the insulating box (extraneous conductors) are permitted.

Third, we can use 'safety extra low voltage' or SELV, a wonderfully bureaucratic term which implies voltages of less than 25V a.c. r.m.s. or 60V d.c. To be classed as SELV equipment, your system must be powered from a battery or from a mains supply incorporating a safety isolating transformer.

Which of these methods you choose for safely powering your own system will depend upon whether you are using battery or mains power and whether you are using a metal enclosure or not. There are other ways of protecting people from dangerous voltages, but these are not really applicable to small equipment. Why not take a look inside existing commercial products to see how it's done.

While on the safety subject, there are just a few more points. Do not have socket outlets which could be mated with cables carrying inappropriate voltages. Above all, reserve the IEC standard mains connectors for just that: mains. Make sure that *sockets* are *live* and that plugs are dead until connected (so that no-one can touch live pins). This makes sense even for low-voltage connections. Segregate mains wiring and low voltage wiring wherever possible. If you're going to depart from the norm in some way, and you're using mains power, check out the regulations to see that what you're doing is wholesome.

Think about what your equipment is going to be connected to as well. Use shrouded connectors to prevent the operator from contacting anything metallic which is part of the circuitry if you want to be really safe, as well as using the precautions outlined above.

Zero volts digital and analogue . . . why the difference?

We've already talked about earth loops above; now we can investigate power supply wiring in more detail. If any piece of wire can be said to be more important that any other to the proper functioning of a circuit, it's the zero volt line.

The common zero-volts connection has different functions as applied to digital and as applied to analogue circuitry. Digital devices think of zero volts as a power supply connection and power supply currents travel along it. A digital zero-volt line is often noisy with current transients pushing hash onto it as the digital i.c.s switch and change state. On the other hand, analogue circuitry regards zero volts as a voltage reference, not usually a supply.

Connecting analogue reference points (the analogue zero volts of a DAC, the non-inverting input of an op-amp) to the digital zero-volt line is a recipe for irredeemable noise. Once digital hash has gotten into an analogue circuit, there's little that will remove it apart from some heavy filtering - and that might also remove some of the information you want. ADCs are a little more resistant than most other analogue circuits; if the noise is worth less than one least significant bit, then it may not show up. But don't bank on it.

Some ADCs and DACs have separate zero volts for this reason; the digital zero-volt pin carries supply currents whereas the analogue zero volts is a reference point for the analogue signal. They both go back to the common zero volts at the power supply eventually, but they should not join together until they actually arrive there. It can be a good idea, where the i.c. is far from the power supply, to put a clamp circuit consisting of two diodes back to back between the two zero-volt pins, to prevent them from straying too far apart.

Occasionally you might find yourself adding an analogue sub-system to an existing digital computer. You have no choice, then, but to connect the two zero volt pins together directly at the chip; it's very unlikely that the computer will have a quiet analogue zero volts, so you don't gain anything. As environments for housing analogue subsystems, computers are actually pretty bad.

I have a digital synthesizer at home which I suspect suffers from mixed zero-volt syndrome. No matter where the volume control is, there is a continual slight hiss of white noise at the same volume. I imagine that it's digital noise due to the zero-volt end of the pot being tied into the digital ground. Some day I'll get around to unhooking the cold end of that pot and I'll route it to somewhere quieter. Well, it's worth a try, anyway.

When is a wire not a wire?

Answer: when it's an inductor or a resistor.

Until room-temperature super-conductors become reality, the conductors of common experience will have resistance. In so far as power supplies are concerned, the resistance of a conductor is its most important parameter, as voltage drops may not be tolerable between the supply and the powered circuit. Current carrying capacity is related to a conductor's resistance by the fact that power will be dissipated in the conductor and it will heat up, but voltage drops may be a concern for long wires before ever their actual current capacity becomes important.

Conductor and resistor are actually words for the same thing. We use the term 'conductor' if we are expecting the article in question to exhibit a small resistance, hopefully as near zero as we can practically manage. We use the word 'resistor' in the fond hope that the resistance of the given article is its dominant feature, rather than, say, its inductance.

To quantify the problem, let's take a nice piece of 7/0.2 wire and run it for 50m to supply a circuit which takes one amp. (7/0.2 wire can take up to 1.4A unless bundled into a cable, when it must be derated to 1A.) We have a ten-volt supply at one end of the wire and our circuit expects to use 10V.

Unfortunately for our well laid plans, the return path of the wire (100m) has a resistance of at least 8Ω, meaning that only 560mA flows in the circuit, and it has less than six volts across it. More than four volts has been lost in the cable.

The stark reality is that, if we want the voltage losses in the cable to be 10% or less, then a wiring resistance of less than an ohm is needed. A cross sectional area (c.s.a.) eight times that of 7/0.2 is needed, i.e. at least 1.75mm^2. The closest size that would suit is 50/0.25 gauge wiring, at 2.5mm^2, 30/0.25 being too small at only 1.5mm^2. And all for a puny one amp.

All this is compounded when we start wanting tens of amps at low voltages (typical of computer power supplies, for instance). The only solution is to work out your current requirements and see whether or not the voltage drops are acceptable. Given time, those situations which merit a closer look will become familiar to you.

The inductances in wiring are a problem too. 100m of 7/0.2 has an inductance of nearly 250 μH. Now with d.c. this is barely a problem, but when a digital i.c. changes its state and decides to pull down the input of another i.c., we may have a change of current one milliamp occurring over a time scale of possibly five nanoseconds. Such a rate of change of current in an inductor of this size would cause a transient of 50V, which is a silly answer. What actually happens is that the switching characteristic of the i.c. is impossibly degraded and the fall time of the output will be stretched, unless the circuit otherwise fails to work due to any other effects of bad power supply regulation.

If you do have a long wire, keep the outgoing and return paths as close together as possible. Decoupling capacitors will also help to defeat this problem. Every so often, every few i.c.s for digital systems in fact, there should be a decoupling capacitor on any circuit board. Make this a low-inductance type (ceramic seems to be the preference with polyester an inexpensive close second) and design p.c.b.s so that these capacitors are as close to the relevant chip as is reasonable.

The effects of inductance and so on are discussed *vis à vis* signals in the section on wiring in Chapter 11. Incidentally, you will not reduce the inductance of a wire by more than 20% even if the wire is made ten times the original diameter; length is by far the biggest factor.

Figure 8.2 shows an ideal power supply layout and interconnection scheme for a small instrument.

Power supply lines should wherever possible run like trees rather than buses or loops. Unless there are good reasons for having a loop (e.g. the integrity of the line is important and there is considered to be redundancy in a loop) then don't use them.

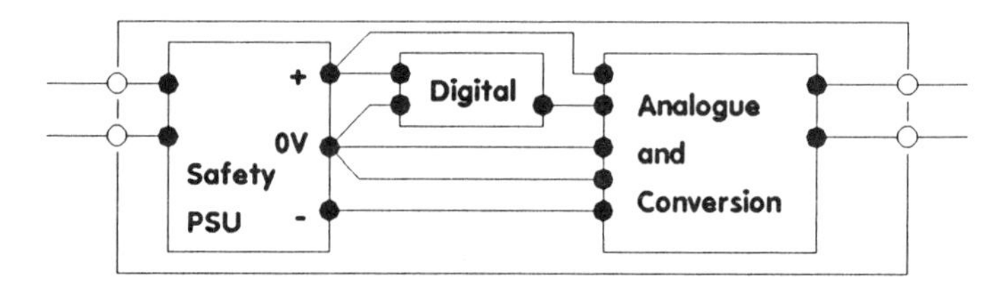

Figure 8.2: Ideal power supply and interconnection scheme.

Buses are so terribly convenient, especially when we get down to board level, that their use is unavoidable really. You can make a kind of 'tree' by putting in a plane which allows currents to flow freely wherever they will and lowers the resistance and inductance of the supply path. However, your interwiring at least can have a tree structure, with each sub-system on a separate branch. At the very least, separating wiring branches for analogue and digital supplies (where these share a common rail) will keep a lot of noise and interference at bay.

You may wonder why Figure 8.2 has two zero volt lines going to the analogue section. One of these is a digital zero volts and one is an analogue zero volts.

Where wiring is lengthy and currents high, you need bigger wire gauges. As a matter of principle use a good size of wire. Unless it's going to be desperately expensive, err on the large side for wiring gauges. Find ways of minimizing the lengths of runs of wire and keep heavy currents close to where they're needed.

The spike, the glitch and the undefined logic level

Thermal noise is due to device physics, and can only be reduced by choosing a component with a lower effective impedance, changing your circuit configuration to allow for that if needs be. Excess noise is due to component construction and unless you make your own components (!) there is very little you can do here except to select a low noise type. Interference is often due to poor layout and is something over which we have more control. We've already discussed wiring layout as far as power supplies are concerned.

We haven't got complete control, however. Mains-borne r.f. interference (r.f.i.) can be injected into our equipment just as easily as it can be caused by our equipment.

I don't propose to say very much about r.f.i., except to give you some general pointers; this is a whole discipline on its own. R.f.i. can arise through actual use of r.f. itself or as a by-product of fast digital systems. Luckily, we can screen or filter r.f.i. Better yet, some circuits will not respond to r.f., being so sluggish that they act as their own filters. But beware of rectification effects here, which give rise to the well-known 'taxi in my stereo' syndrome.

The thing about r.f.i., if it is going to be a problem for you, is not to let it into the box in the first place. Once inside, it can radiate to other places and cause problems. Mains filters prevent other people's r.f. pollution from getting into your instruments and also do a good job of preventing any hash that you're generating from getting back into the mains.

The filter shunts high frequency interference components to earth via a capacitor and prevents the escape (or ingress) of r.f.i. onto the mains lead by using in-line inductors. This does need an earth. Such units can be obtained off the shelf.

Screening helps, of course. Metal boxes need to be earthed otherwise they'll just re-radiate the r.f. Try not to use a cable screen as a signal return, use twin screened where possible, otherwise signal return currents will flow in the braid and couple with other things (this applies equally to the prevention of hum in audio systems where a braid may carry 50Hz induction current).

Try turning circuits around or placing capacitors at various points if you need to. Anything to outwit that pesky interference!

The transient or 'spike' (so called due to its shape if viewed on an oscilloscope) is a special case of noise, arising from an impulse typically on a mains supply or arriving along a long line, having been induced by a lightning strike, changing taps in the power company's sub-stations or other such event. Spikes can appear on the mains power line or on any long line used for communication between remote sites.

Protection from spikes involves putting a transient suppressor across the line to be protected. These shunt the energy of the impulse away from your nice sensitive circuitry. Provided no permanent damage is done, analogue systems will recover from transients and will after a time return to their normal working state. With most digital systems, despite their good initial tolerance of noisy environments, if the spike is strong enough to penetrate to the extent that the circuit operation is upset, then programs will crash or data will be corrupted.

Often a transient suppressor and mains filter are incorporated in the same package. For some other examples of transients, see my comments on driving relay coils in Chapter 6.

This brings us neatly on to the subject of glitches. Glitches are a kind of spike, I suppose, which don't exceed the usual voltages for logic levels and which are of our own making due to poor logic design. They are very narrow pulses which will sometimes clock counters when we don't want them clocked, set or reset latches which don't need to be changed, or write odd data to registers at times when it shouldn't happen.

Glitches mostly arise from 'race errors'. Where there are two paths through a series of logic gates, each with a slightly different propagation time, and where the result of these two paths are combined once more, then there is potential for a glitch, especially if the result of changing the inputs gives the same output as before. There may be an instant between the arrival of the one path and the arrival of the other when the output goes to a different state momentarily.

Race errors commonly occur where a counter output is being checked for a particular output value. There may be an instant, as the count ripples through the counter, when the counter output agrees with the set value, and there will be a glitch. Ripple carry counters are prone to this; synchronous counters are not.

If you design logic with a master clock, glitches may become unimportant. That is to say, they will still happen but will not be acted upon because all the latching and counting waits for the edge of the clock, giving time for things to settle properly. Computers rely on this synchronisation heavily.

In a similar vein, tying unused inputs up properly will have beneficial effects on your circuit's performance by preventing noise from coupling into the exposed inputs. In particular, CMOS can never be left floating; the input has such a high impedance that noise of any kind, even mains hum, will couple to it quite nicely. TTL can be left to float HIGH, even so I don't like it, please pull unused inputs up properly. LS TTL, like CMOS, can be connected straight to the positive rail, but standard TTL must use pull-up resistors (values of 1k0Ω to 4k7Ω are usual). All TTL and CMOS logic families can be tied straight to zero volts for a LOW.

Related to the above, undefined logic levels are less of a problem. They tend to arise where a poor job has been made of an interface between logic families. There's not a great need for mixed logic families; avoid it where possible. I did see an example once where a CMOS chip was tied to a TTL chip. It worked fine if the CMOS chip was a Fairchild type, but just you try a Motorola sample, buddy. Not the fault of the i.c. manufacturers, I hasten to add.

Where a voltage is between the guaranteed maximum LOW V_{IL} and the guaranteed minimum HIGH V_{IH}, it could be interpreted as anything. Logic signals should pass through this region without lingering, unless you plan to use a Schmitt trigger input gate.

9: Design for Reliability and Serviceability

Every part of this book, really, has had something to do with reliability, since that permeates the whole of the design and manufacturing effort. Here I will be looking a little more closely, however, at those things which we can do to especially to guarantee reliability, along with other desirable features such as adjustability, usability, maintainability and constructibility.

Designing for ruggedness

Ruggedness might imply electronic ruggedness, physical ruggedness, or both. Usually when the term rugged is used there is a degree of both implied. Ruggedness might also imply that a system carries on working to specification under environmental extremes (remember that incoming signal and external supplies are part of the environment) or it may just imply that the thing doesn't break, even if its operation is impaired whilst the extreme condition is being experienced.

A typical grading scheme might run from the least rugged to the most rugged along the following lines: domestic, commercial, industrial, military. It may be useful to classify the ruggedness of your own product, however loosely, on one of the above categories.

To take physical ruggedness first. We need to decide who will be using the system and where. This should be sorted out at the specification stage; if not, then it needs to be thought upon soon.

If the system is to be used outdoors or in some extreme environment, then the packaging will dominate the design to a large extent. In some cases, there may not be an enclosure available to suit, in which case you may need to make up your own.

Physical ruggedness often implies that the best grade of materials has been used and that expensive connectors and other furniture have been used. This is most likely, but not necessarily, the case. It can be that by cleverly arranging the likes of carrying handles and other furniture, or by having recessed panels for example, that fairly ordinary connectors and indicators can be well protected, certainly from the point of view of damage due to dropping or guillotine-type accidents.

Even where a connector is of the best grade, there is no harm in recessing it or otherwise protecting it further. When recessing, though, be sure that the operator can still reach a connector with enough grip to make or break the connection.

Think about dust, water and vibration. These are the physical enemies of any electronics.

The Index of Protection (IP) rating deserves an explanation here. Essentially, the first number of the rating relates to the ingress of solids and the second number relates to the ingress of water. The higher the number, the better the protection. Zero means entirely unprotected.

The least protection which usually gets a mention is IP20, which means that you can't poke your finger into the enclosure but there's no protection from dampness. IP20 is typical of, say, domestic audio equipment.

Protection to IP33 would mean that wires more than 2.5mm diameter cannot enter the enclosure and that rain falling at an angle of 60° or less to the vertical will have no harmful effects.

The best possible protection is IP68, which implies that the enclosure can be submerged in water for extended periods and that fine talc finds no entry.

Enclosures to IP56 or IP65 are commonly available and are quite satisfactory for general outdoor use, although a wee roof, lean-to or other shelter to prevent direct driving rain is often not a bad plan. Other wrinkles are to put cable glands in the bottom of the enclosure where convenient, as they will stay drier that way, and to leave a loop (or 'bight' for the seamen amongst us) in any cables which will allow any water which does get onto the cable to drip off rather than run into the gland.

This latter is a favoured trick of those who must drill through window frames to put in television aerial downleads and who dislike the idea of rainwater running down the cable and straight in through any tiny gaps in the seal.

Remember not to compromise the protection rating of the enclosure by fitting furniture (switches, indicators, cable glands) of a lesser rating than the enclosure itself. The protection rating of the system as a whole will be that of the least well protected point, on the principle of the weakest link in the chain. For the same reasons, use the external lugs when mounting such an enclosure; don't go drilling holes in the back to mount it!

As far as vibration is concerned, there are two things which could happen. Things can work loose (nuts, bolts and so on) or they can fatigue and eventually snap (component leads, wiring).

To prevent slackening, you can use washers, anaerobic adhesives (locking compounds) or special locking nuts.

All kinds of washers can help prevent slackening of the fastening due to vibration, particularly single coil (or 'spring') washers and star ('shakeproof') washers. Their manifold uses are described at greater length in Chapter 11.

Locking compounds are proof against vibration but are intended to release the joint when deliberately undone; if you ever want to get fasteners undone again, don't use an ordinary adhesive. A small tube of this compound can be bought at any garage as well as the more usual electrical suppliers. It is quite runny and it can emerge from the tube very quickly if you're not careful; you only need to use a dot. Nail polish is also used to lock the screws of trimmer potentiometers. Red is the preferred colour; but nowadays pots are more vibration resistant anyway, I am told.

Split or misshapen nuts or nuts with nylon locking inserts which grab the threads of the bolt are also used to prevent slackening. Other techniques, such as the cotter pin and the castellated nut and its relatives, really belong in the world of heavy engineering. Note, it is not necessary to use a nylon locking nut *and* a locking compound at the same time.

High accelerations (due to a hard landing) could cause shearing of bolts if there is any heavy equipment mounted inside the enclosure, as well as breaking the enclosure itself, of course. Mount heavy items flat to panels or to the mounting plate rather than on pillars, to minimize leverage.

Extreme physical ruggedness can be had by shock-mounting electronics inside an extremely tough outer shell.

Car designers are experts in anti-vibration mounting. Learn from poking around inside a car engine compartment.

Fatigue of components can be combated by potting or by 'dead-bugging'. Potting can be done either by pouring compound around the circuit in a potting box, or by smearing something a little stiffer (silicone rubber) over the finished board. The latter is easier to repair. You cannot pot anything which is going to get hot and which needs airflow around it.

Dead-bugging refers to the technique of putting bends in the leads of components to give a bend, about 5mm tall, which sticks up away from the circuit board. I suppose the name derives from the fact that the poor component now has its legs sticking in the air like a deceased caterpillar. This removes any pull that the board may exert on the component (and vice versa) and is particularly good against the rigours of thermal cycling as well as vibration. It is often used along with the 'smear'-type potting.

The prevention of fatigue in wiring has much to do with good assembly practice, which I will be exploring at greater length in Chapter 11. From the point of view of the designer, what needs to be done is to specify connectors and cables which are intended for each other, such that crimps mate securely and strain relief arrangements grip the cable properly.

The other factor is to see that cables and wiring looms are fastened securely and do not pull on the connector or the boards. My favourite means of fastening looms is the screw-mounting saddle which accommodates a cable tie; it has a positive grip about it. I don't like self-adhesive cable grips; they're tacky, in more ways than one, and will soon give up the ghost if called upon to do any real work, particularly those with the soft aluminium fingers.

Leave room for a little slack; a tidy loop of wire about 2cm in diameter adjacent to a barrier strip or similar looks good and acts as a strain relief rather in the manner of the dead-bugging mentioned above. By the way, avoid 'choc blocks' as connectors unless they have a wire protection leaf. They have a habit of slackening with time. Use a proper barrier strip instead. The same goes for the cheaper three-pin mains plugs. Unless you are going to use a ready made or assembled mains lead, try and get hold of plugs which have the nut with the captive washer. MK make them; they're the best.

Electrical ruggedness is concerned with two areas: power and signals. Systems which can operate from a number of power sources and which can withstand overloads on the signal lines may be considered rugged.

The trend nowadays is to provide power supplies which can accommodate a number of different input voltages. Certainly, I know of marine and aviation electronics which can function quite happily on anything from a practically dead 12V battery to a fully charged 24V system, not always with a change of input tapping. There is a similar trend in a.c. supplies too; there are systems which will supply stable d.c. from either 110V mains or 240V mains, with no manual switchover. This is a real boon; no more blown up supplies due to incorrectly set voltage selectors.

Always fuse inlets properly and label power inlets and fuses with voltages and current capacity.

The comments we can make about ruggedizing signal connections apply generally to power too. We need to guard against improper use (people plugging in the wrong cable) and electrical events or accidents (lightning strike, EMP). Make connectors unique where possible. Distinguish firmly between a.c. mains connectors and all others. (It is permissible to carry low-voltage a.c. or d.c. in the same connector as certain signals.)

Where connectors are not unique, key them to prevent incorrect insertion (this is done by inserting a plastic blanking pin into the socket half, along with removal of the corresponding plug pin, and is not possible for all connectors) or label them. Where it isn't possible to key or label connectors, try to make them interchangeable by arranging the pin allocations so that they either still function or that no harm is done by incorrect insertion.

I remember hunting high and low for a fault on a gadget intended to plug into the parallel port on a laptop computer. Imagine my chagrin, then, when I discovered that precisely the same connector had been used to provide for the external floppy drive; we'd duly connected our wonderful widget to the floppy controller. Thankfully, both laptop and widget lived to tell the tale.

Again, label connectors with their function or purpose. (The connectors on the laptop above were labelled using a raised moulding, but we couldn't read them in the confined space in which we were working.)

Circuit-wise, design with the worst case in mind. For example, design an RS232 input which will also pick up normal TTL logic levels, as not every maker of equipment with RS232 outputs bothers to provide an output which swings both ways about zero volts. It's the easiest thing in the world to make an RS232 input which switches at +1.5V, say, rather than zero volts.

Barrier protection can provide increased ruggedness. A little extra circuitry to limit input voltages, especially on those systems where the user is invited to plug in their own equipment, is important. A naked CMOS input, for example, tied straight to an external connector is asking to be blown; a series resistor and clamping diodes will prevent this happening.

Better yet is a transistor input stage which will pick up reasonable signals but which will withstand some maltreatment under overload conditions. Consider optical isolation as a real option for any digital signals which have to travel more than a metre or so, especially in electrically noisy environments. At the very least, for this example, use a pull-up or pull-down resistor which puts that circuit function into a disabled state until the connector is inserted.

So far as outputs are concerned, fusing or current limiting is a good plan.

Ruggedness in software is a whole other ball game. The software is as bug-free as we can make it and runs fine, but how will it come back up after an unscheduled power down, and can it recover from a corruption or crash? I don't want to spend a long time on this subject but there are a couple of things which will get you out of a deal of trouble.

First, don't put watchdog circuitry to an interrupt line. An interrupt service routine can run merrily, aeons after the main program has crashed. I have personal experience of that! Arrange watchdog software in the main program loop.

Second, fill unused areas of your ROMs with NOP (no-operation) instructions, if your chosen microprocessor has such. If the program crashes then there's a good chance of it landing up in one of these areas. If it does, it will synchronize itself; put a restart or jump at the end of the NOPs to get the program going again.

Third, use an error checking protocol (a cyclic redundancy check is probably the best) with re-transmission of faulty data frames. This is generally only possible when you're directly responsible for both halves of the communicating system and you can dictate the protocol.

Incidentally, interrupts are reputed to be a source of unreliability. I would put it slightly differently and say, rather, that they are a source of *uncertainty*. Those people who must use formal ways of designing software (since they need to formally verify that their software is correct) abhor interrupts, since there is no way yet of predicting the precise actions of interrupts. Statistical methods must be used instead. Indeed, there have been processor designs, notably the Viper, which have no interrupts, precisely for these reasons.

But you can stay out of trouble with interrupts by using an old trick, i.e. buffers. Isolate any communications by using buffers. Don't feed information direct to your program, but get the interrupt service routine to feed the buffer. The service routine will have to be a little bit intelligent, and refuse to transmit the data from an empty transmit buffer or to put further data in an already full receive buffer.

Your main program feeds data into the buffer and leaves it to the transmit service routine to send it, after having set it going by calling it once explicitly. Your main program can also pick up data from the receive buffer at leisure. It is important to get the main program to temporarily disable the interrupts before it messes with any of the buffers itself. The main program and service routines talk to each other by setting up pointers which tell each whether the buffer is empty, or where the next chunk of data is when the time comes to get it. Thus, synchronization is obtained. These techniques are appropriate for data streams; you don't need data buffers for simple bitwise control of the likes of lamps, solenoids and motors.

There should either be plenty of room in the buffers and plenty of processor time in which to process incoming data, and there should be some way to 'turn off the tap' for received data if you are not to lose any. If you have used all these precautions then your interrupt scheme will be watertight, even if there is still some uncertainty about exactly what is going to happen when.

My own experiments in the areas of redundancy checking and multi-level interrupts led to a system in which a pair of master and slave units, normally in constant communication, could each be switched on or off at random with no loss of data. Very robust.

Designing for adjustment

Adjustment has to do with trimming out errors and allowing the operator control of the system. As such, it has to be done either at setting-up time, 'with the covers off', or on a moment by moment basis by the operator during normal operation.

There may be some problem in choosing between these adjustment strategies. One the one hand, keeping everything buried inside the enclosure results in simplified day-to-day operation, with less likelihood of mistakes, but does not allow the operator the ultimate in control or adjustability.

On the other hand, having every possible control to hand can be confusing, especially if it relates to some purely internal function whose effect is not directly and obviously noticeable in the behaviour of the system as a whole. For this reason, it's often a good idea to provide some indication of what effect an adjustment is having or, at least, to provide easily located test points such that instruments are more easily attached.

One strategy to employ is to make adjustment easy but not so easy that it can be done from the front panel. In other words, there may be a segregated panel which can be opened to reveal controls which can be easily adjusted by a trained operator without dismantling the system. In fact, it is bad planning to make it necessary to take something to pieces completely to adjust it, although it does happen; don't worry, I've been guilty of that in the past, too.

The other big question about adjustability is whether or not something should be adjustable *at all*. It's perfectly possible to use particular circuit configurations which are insensitive to certain errors or to specify components to the kind of tolerance where adjustment is not needed. This results in a physically more robust design too. Think hard about the accuracy demanded from your system; if it seems that an adjustment is not needed, then avoid using it.

Avoid designs, though, which avoid the need for adjustment by requiring matched or selected components, the so-called 'adjust on test'. Selection is a fiddly job which just adds to the cost of production. Any adjustment is an easier prospect to face than eternally ploughing through a set of resistors or capacitors (one is reminded of the Flying Dutchman) trying to find the one which just cancels out some error or other.

Offset nulling of operational amplifiers is fine if you need it. Often you can get away with no offset null in a circuit and then at the last gasp have a zeroing pot which renders the 'final account' and gets rid of any accumulated errors in one fell swoop. For instance, a filter with several stages can have its zero level shifted outside the filter itself, before or after the filter in fact. Unless you need the outputs from the other points in the circuit, one adjustment is preferable to several. Keep it simple!

To get down to the actual component values, then. Most adjustments you will specify will be potentiometers which will form part of a potential divider network. You need to decide how much current you want to flow in the potential divider, which will determine the overall resistance. Then you need to decide how much adjustment is needed.

How coarse should the adjustment be? Too coarse an adjustment can be difficult to set; too fine, and it may not be able to bring the circuit into alignment. Ideally, the adjustment should be just sufficiently coarse to account for tolerances in the circuitry with a little margin, 20% say, to play with. Limit the amount of adjustment by using padding resistors at each end of the pot.

There are two ways of looking at the calculations here. One is to focus on the requirement that the centre of the pot track shall carry so-and-so volts, and that the ends are so-many percent distant from that voltage. The other, perhaps simpler, way is to specify that there are so-many volts at the top of the adjustment potentiometer and so-many at the bottom.

In either case, simple application of $V = IR$ should give you some component values. Then choose standard component values which approximate to the ones your calculator gives you, then check that the chosen standard values really give acceptable results. (I have explained the use of spreadsheets for this little job in Chapter 10.)

Be careful that, when your carefully calculated adjustment is actually in place, loading due to currents drawn by the rest of the circuit does not render all your calculations worthless. Try to use circuit configurations in which potential dividers are not loaded by the rest of the circuit, particularly when the adjustment in question has a nice graduated scale attached! Loading which might vary due to other changes or adjustments is an evil prospect; avoid that fate at all costs.

Designing for maintainability

Adjustment is part of maintainability. Ease of adjustment goes a long way to making for good maintainability, but in this section I'll look briefly at some other issues such as access and the availability of parts.

Ideally, it ought to be that one can get at any part of a system without removing any other part. I have worked on systems where the machine had to be literally pulled in half before the cover could be removed from the one side, although I suspect this is a rare example. It's not always possible to get directly to an offending part, but please avoid these extremes.

Train people in the ways of gaining access, or explain it in the user guide or service guide. I have a couple of amusing tales about that.

An ex-colleague of mine was once extensively trained in the arts of fixing the internals of printers. I daresay that he could have drawn the circuit diagrams from memory. Anyway, they failed to tell him *how to get the covers off*. On his first job he had to suffer the embarrassment of abandoning the afflicted machine while he enquired how this was to be done.

On my own account, I once had to fit a communications option to a printer. No, they had lost the manual. Now all extra cards go inside printers, do they not? So I spent a good part of a morning wondering where on earth this thing was supposed to plug into. It was obvious to everyone else (apparently) that the card went on the *outside*, under a little cover which popped off. Such is life.

The moral is, tell people the obvious things too. Use a little psychology and don't be too obvious in the way you say it, that's all.

How easy is it to replace parts in your design? There are two issues here really: first, can you then fit them easily and second, can you actually get them.

Access has a great deal to do with the former, but there are additional considerations. Board replacement has become a popular means of fitting replacement parts, with the dead boards usually being returned to base for repair to component level. Make sure that boards are easily removable, either by using standard bus schemes or by making wiring looms easy to detach.

If field personnel must work to component level, then socketing i.c.s is a good plan, although this is going out of the window nowadays too, with the advent of surface mount components.

From the point of view of availability of components, there are a few things which could be done to make sure that this runs smoothly. Where a manufacturer suggests that a component should not be used for new designs, this is a sure sign that it will not be as easily available as a replacement part in future. Keep a few spare components and boards on tap, ready to be used as replacements. Issue a small spares kit with your systems which contains operator-replaceable parts such as bulbs, fuses and so forth.

Availability of information. Make sure that the people who are to work on the system are clued up and that they have access to the information they need to make the job go well.

Does your design need special tools for access or adjustment? There is a fine line between restricting access for safety and to prevent tampering and restricting access to maintain a monopoly on service call-outs. Restricting access (or information) to maintain monopoly is a commercial decision, not a technical one. I always avoid the need for special tools unless it is forced upon me by the nature of the design.

Finally, avoidance of brinkmanship in design will assist maintainability. After all, if your product only just works on the bench, then how is some poor service engineer going to cope with it out in the field?

Designing for usability

Some of the comments made above, concerning designing for ruggedness and adjustment, overlap with concerns about usability.

The ability to modify the system for other purposes is one aspect of usability; the degree of specialization of an instrument, for example, has a lot to do with whether or not it can be turned to some other purpose. The other aspect of usability is how easily the operators can fulfil their chosen aims without other functions of the system intruding.

A big part of usability has its roots in cognitive psychology. There are some essential questions to be asked here:

Can the operator see, at a glance, the information he is looking for, or must it be searched for?

Is there any ambiguity in layout or labelling which might cause one piece of information to be mistaken for another?

How well labelled are the controls of a system and do they mean what they say?

Ambiguous or inadequate labelling is a curse. Finding good expressive labelling can be a bit of an art form, but when its done well it's really worthwhile. Consider using symbols, icons if you like, as well as text.

The means of operating the device should be as clear as possible, with complex operating procedures kept to a minimum. In a word, an uncluttered aspect, both operationally and in the layout, makes for good usability.

Designing for constructibility

Constructibility refers to our ability to easily assemble the system with no need to use awkward or time-consuming processes. Adjustability is an issue here too, the initial adjustment being an integral part of manufacture. It's too easy to design something which cannot be built. There's an expression for a design like that: 'manifestly incomplete'.

Designers should be conversant with the materials and components they use. They must also be able to shape or assemble these things in a way which is elegant and easy to do. As far as the actual electronics goes, it is mostly sufficient to know where to look and to be able to interpret the figures and data given.

Particularly as regards the mechanical or physical side, we come back to the importance of visualizing the system in the act of being made. On several occasions I've seen components (panels and power supplies in particular) which obscure a screw which itself holds the part in place! Talk about chickens and eggs.

Use your imagination and try to see your own hands actually getting inside the box and doing the job. There would be far fewer poor designs and far less need for special tools of designers did that just now and again. Or indeed if they built or used their own designs.

One thing which often happens is that designers forget that screwdrivers are finite in length rather than of zero length. Only perfect screwdrivers are of zero length. Often, the screwdriver fouls the opposite side of the box when trying to grapple a screw head. Keep a stubby screwdriver in your kit, or perhaps the cranked variety. Better yet, design for good access.

If you have a CAD system which allows you to drag shapes, make up some shapes of the larger components and drag them around inside a rectangle which represents your enclosure. Again, it's kindergarten stuff, but there's absolutely nothing wrong with it.

Again, the issue of availability of parts rears its head. Some manufacturers are bad for discontinuing parts without notice; try to use parts which are likely to be around for a time.

10: Tools for Construction

This chapter is about small hand tools; if you like, the strippers, grippers, nippers, clippers, nibblers and screwdrivers of this world. I've also included some information on workshop practice. I haven't described every possible tool here, there are more kinds than you can shake a stick at, I've just given a general guide to the most commonly useful, along with some tips as to their effective use.

The voltmeter, oscilloscope and other instruments have been described in Chapter 4. Little more needs to be said about them at this point except for the need to keep them handy, especially if you're using test-while-construct, i.e. stripboard or similar. P.c.b.s are much less demanding of testing during construction since all the errors have been discovered, we hope, at the p.c.b. layout stage and there is less opportunity for odd mistakes to creep in.

Small tool storage

First of all, where to keep tools. I find one of the little plastic sewing boxes with the lift out tray is fine for small hand tools. Alas, the handle on mine is due to split any day now (indeed it has done, since I wrote the first draft) so I'll need a new one unless I apply the sticking plaster liberally.

The only disadvantage with this particular kind of toolbox is that when you forget to latch the lid before picking the box up, you get a kind of tool fricassée on the workshop floor. Try and get one with proper hinges with a pin rather than one with a moulded plastic hinge. If you've got lots of bigger or heavier tools then a steel toolbox is not an expensive item. But they do tend to get scratched and rusty.

You might also think about a tool board, with the silhouettes of the tools drawn onto a plywood or chipboard sheet and nails or hooks proved to hang them up, if you have loads of wall space and if portability isn't an issue for you.

Wiring

It is essential to have a really good pair of snips. For close electronics work, get the flush cutting kind rather than the heavy electrician's bevelled edge jobs (although a pair of these might be handy, too).

Every so often a set of snips comes along whose performance belies their price; I'm thinking in particular of a kind made from steel plate bent, riveted and sharpened up which really out-performs anything else at the price.

Do not cut anything but copper wire and component legs with your good pair of snips. Use an old, gash pair of snips for any rough stuff. Cutting some materials ruins the cutting edges by nicking them. Holding a pair of snips up to the light with the blades closed will reveal any such tiny dents. A pair of snips misused in this way is a sorry sight. The other side of the coin is, I suppose, an inferior pair of snips could be damaged by cutting ordinary copper wire. Make your supplier's life tough if this happens to you.

My Dad used his teeth as wire strippers. I suppose we Cuthbertsons have strong teeth but even so it's not recommended.

There are some dreadful strippers and some excellent ones. The pistol grip type can be surprisingly good although a bit chancy. A favourite of mine is the kind with a series of graduated holes. There is no fiddling and it doesn't go out of adjustment. The type which looks like a praying mantis, which you've to pull, I've always found awkward. The ones with the tiny wheels which you turn never seem to be in the same place twice and you're always tweaking them instead of stripping wires.

There are co-axial ones which are OK if you're doing a lot of co-ax; the handiest type cut everything to a well defined length as well as just simply stripping. Otherwise use a penknife or craft knife for the outer, just like any good old electrician does with his twin-and-earth.

As well as being useful in all manner of places, a craft knife can be used for stripping the outer insulation from cables. One tip is not to use a sawing action with the knife, as this will cut the insulation of the cores underneath and may even part any braid. Rather, roll the cable along the blade, lifting away and bending the cable away from the cut each time so that the insulation splits without damaging what's inside.

Amongst the most expensive hand tools are the crimping tools. You can pay a hundred pounds or more for some types. If you find any exotic looking sets in a jumble sale and they don't look worn, grab them first and ask questions later.

Crimping connectors saves time, but it often isn't up to the standard of a good solder joint. Things can go wrong with a crimp that are not obvious and if an unsuccessful crimp is made then that renders some kinds of connectors unusable; you don't get a second chance. For preference, take the time to learn how to solder such things as BNCs to make a proper job.

BNCs and the discretely-pinned Dee types are in my experience the worst crimpers. There are some kinds of crimps, on the other hand, which always seem to work. I refer in particular to the KK types which are probably one of the neatest ways of making discrete, detachable wiring up to a board. Also the QM types seem to crimp reliably, as do the discrete insulated tags of the kind often found in vehicles. Oddly, all these types have the least expensive crimping tools . . ..

Remember that the insulation on a wire needs to be included in the crimp terminal. Usually, you need to crimp twice: once to secure the wire core to the terminal, and once to crimp the insulation into the strain relief part of the terminal.

If you really develop a taste for crimping you will probably need a few dozen tools as there are as many kinds of crimps and standards are few. While on the subject of non-soldering connectors, there is one kind that makes life very easy. Not only that, but this kind does not need special tools to assemble, although there are firms which will sell you a special jig anyway. I'm talking about insulation displacement connectors (IDCs) here, the kind which are used with ribbon cable.

To use IDC connectors, first chuck away the plastic jig things that you bought at vast expense. Invest in a pair of 'soft jaws' for your vice instead, or fix up your own by shaping and bending an aluminium plate to slide over each jaw of the vice and sticking a stiff rubber pad to the face (these will come in handy for other uses too). You must use soft jaws for this as the serrated faces of an ordinary vice do cruel and woeful things to connectors.

Pre-assemble the two halves of the connector with the ribbon between the two halves (for many connectors this is the 'first click' of the latching lugs). Turn the vice until the ends of the connector just click into place. Any further and you may damage the connector by compressing the cable too far and possibly snapping the lugs which latch the two halves together.

If this connector is the end of a cable (the usual case) and if you have not used a pair of shears to cut the ribbon, use a sharp craft knife to trim off any excess fragments of cable sticking out of the connector and then check that there are no whiskers of wire shorting adjacent cores. Finally, any strain relief moulding should now just clip over the doubled-back ribbon cable. I cannot remember ever having a connector fail on me if done in this way.

Sheet metal

Nibblers can be a useful tool if you're doing a lot of metal work and need to make odd-shaped holes in thin panels. A nibbler looks rather like a pair of pliers. After drilling a small starting hole in the middle of the hole you want to create, you use the jaws of the nibbler to 'nibble' away at the edges to give you the shape you need.

Nibblers are not the cheapest tools and I've seen them destroyed by attempting to cut material which is too thick or too tough to handle. Use them only on the recommended materials and thicknesses. There is a range of nibbling tools (square, ovoid, etc.) available for bench presses too; such nibbling tools are probably rather more robust than the hand-held 'pliers' type.

While on the subject of holes in panels, the chassis cutters can be handy if you're needing neat holes fairly quickly. The square ones can get chewed up at the corners. Don't bother with small round sizes; use a drill instead.

If you're plagued with burring of the holes on larger drill sizes, there's a trick a chap showed me recently. It involves grinding the end of the larger sizes of drill to leave a central spike and two 'wings'. The spike keeps the drill true while the wings cut a neat hole in the panel, rather in the manner of a tank cutter. If your grinding is good, a neat disc of metal will fall out of the panel, leaving a good edge.

Think about getting a tapered hole cutter, step cutter or similar device for larger holes too. These tend to leave a clean edge to the hole. A reamer, preferably with a T handle, can be an asset for deburring and adjusting medium-sized holes. A spare drill chuck can also be used to hold a tapered cutter, to the same effect as a reamer.

When you graduate to full-scale production, a bench press with the appropriate dies and punches is a distinct possibility. The price of die and punch sets will make you gasp but they are a real time-saver (thirty seconds for a complex hole shape as opposed to thirty minutes). If you need to cut a few hundred holes all of the same size and shape then a bench press is the thing to go for. Alternatively, farm the job out.

A set of files is a necessity. Until you get a nice bench press, you'll be using files to shape all kinds of holes. Even after you've got the press, you'll still need files occasionally. A set of instrument files is cheap enough; a larger flat file (about an inch across) is useful, as is a round file about a quarter inch in diameter.

I am eternally grateful to the mechanic who took me to one side while I was a young loon at Edinburgh and taught me how to use a file properly. 'Yon's a cutting tool', he said. Up until that time I'd been doing what most people do with a file, busily swishing back and forth, panel in one hand and file in the other, occasionally resting the panel on the edge of a bench (and risking scratching thereby, incidentally), eventually wearing the metal away to the desired size. He showed me that a single cut done in the right way was worth a good half minute's thrashing. If someone offers you advice of this kind, it's worth taking them seriously.

Put the panel in a vice, sandwiched between two straight pieces of wood. The front piece of wood should be aligned with the straight edge of the hole as a guide. This prevents filing past the edge of the hole and also stops the metal buckling if you're working close to the edge. Grip both ends of the file and use a single long forward stroke to cut the metal; use the whole cutting length of the file. Practice makes perfect, so try it some time.

Aluminium, electronics' favourite metal after copper, is a soft metal which clogs fine files, so use a coarse file when filing aluminium. Oddly, this results in a perfectly good smooth edge. A clogged file will rip and smear the edge you're filing. Use a fine wire brush to remove clogging before it builds up. Since a file is a cutting tool, it can be 'sharp' or 'blunt'. It goes without saying that a sharp file works best. Sharp files clog less easily too.

Keep files clean of grease as well; they will clog very much more easily if there is a trace of old grease on them.

A hacksaw is also a necessity. Use the right kind of blade for metal, note.

You can also get 'rod saws', blades which cut in any direction, and tension files which are very similar. These are very useful for cutting awkwardly shaped holes.

Keep the plastic backing on your aluminium sheet, if it has one, to prevent scratching. Mark, scribe and punch the back of the panel where mistakes will not show.

Soldering

The soldering iron is to hand tools what the oscilloscope is to instruments; it immediately distinguishes the owner as one of those electronics crowd, a breed apart from the rest of humanity.

It's possible that you'll spend as much time soldering as anything else. For most electronics at the chip scale, use a small iron. For larger pieces of metal, a bigger iron is a boon; the smaller tips lose heat rapidly (low thermal mass). Even modestly sized solder tags and 24/0.2 wire may need a larger size than the miniature irons. Soldering to the metal cases of potentiometers, a popular screening ploy in audio equipment, needs a bigger soldering iron too. My own soldering iron is a little 18W job which is fine for about 95% of what I do; I also keep a 'brute' soldering iron which sees occasional service.

After a while bits will corrode. What tends to happen is that a void forms under the face of the bit; a bit in this state will not conduct heat to the workpiece efficiently. Don't soldier on with a worn bit for too long; it can be frustrating. You can give an old bit a new lease of life by filing it down, but this is not really satisfactory as there will be no coating on the tip.

Be careful with soldering irons. Burns are about as common as shocks and only slightly less so than cuts (in my own experience) and we could do without any of them. The additional danger with irons is fire risk as well as personal injury; don't have loose papers flying around near a soldering iron and return it to its holder when it's not in your hand. Melting cable syndrome is something else to look out for. Test leads are notorious for emitting nasty smells as they are draped across a hot soldering iron.

Some tips on good soldering practice are in order.

Keep the sponge supplied with your iron damp and use it frequently to clean excess flux deposits from around the bit. If you haven't used your iron for a minute or two, i.e. it has been sitting in its holder while you've been getting on with something else, just touch the solder to the bit for an instant to wet the bit.

The soldering iron should touch the workpiece and heat up the workpiece before the solder is brought up to the joint to run in. Conversely, the solder should be taken away from the joint before the iron is taken away. Work quickly but allow sufficient time for the solder to flow freely around the joint. Note, the solder must *flow* - we don't gradually work-up layers of solder.

Where the component lead is substantial, heat it before the p.c.b. pad by just running the iron tip down the last quarter-inch.

If you ever need to solder to a tag which is fixed to a metal chassis (an earth connection perhaps), it is not a bad idea to release the tag from the chassis first; in this way, the heat-sinking effect of the chassis is removed, making soldering easier.

Although it's easy to tack a wire onto a spill or terminal, and this is often done as a temporary hook-up for test purposes, a more permanent result is to be had by taking the wire once around the terminal and then letting solder flow into the joint. Solder buckets should be filled with solder and the wire tinned prior to heating the bucket and pushing home the wire.

The finished joint should be bright rather than a dull colour and should have no blow holes in it (these appear like tiny black dots). It should be pyramidal in shape, at least for a joint between a led and a p.c.b. track, not a blob shape.

One last thing. If you ever need three hands while soldering, one to hold the iron, one to hold the workpiece and one to hold the solder, then you could use a little wrinkle that I call the 'snake charmer'. Wind half a metre or so of solder around two fingers, not too tight. Now use the coil as a stand and pull one end of the solder up into the air so that it stands up like the familiar snake in the basket. You should now be able to bring the three things together . . . as the solder wears down, keep pulling a length out every so often.

I've explained soldering in this section of the book rather than in Chapter 12 since it is a fundamental process employed everywhere.

Assembly

Crop and bend tools are not really that useful at this stage, unless you're in a real hurry, i.e. in full production. In fact they make the component lead lie flat to the board or crimp the end of the cut lead, making it nearly impossible to remove the component later without cutting the lead near the component body.

I'll describe an alternative 'bend' you can use instead of using the crop and bend tool. If you pull the component lead through the board with a pair of pliers until the component body is sitting above the board at the required height, then grasp the component body with the other hand and pull sideways (in the plane of the p.c.b.) you get a kink of 10° to 20° which is sufficient to hold the component in place until the solder can be got at it. It removes the need for the foam retaining devices too (these can be a pain if you have a standy-uppy component rather than a prone component).

Small transistors and other tall components can be dealt with in the same way as prone axial components; just don't pull the leads all the way through the board. Apply solder then crop the lead just above the joint.

Get a set of flat tip screwdrivers and cross-points. It's worth getting good screwdrivers. I struggled for years with cheap screwdrivers until a joiner pal of mine put me onto a moderately-priced imported screwdriver. I was working on the boat at the time, trying to get really recalcitrant, corroded screws out of the mast. Since that time, I've never looked back.

A good screwdriver will extract screws which look impossibly tight or which have damaged heads. There's nothing worse really than a soft tip which twists or smears and leaves the head of the screw either untouched (at best) or with a slot with rounded shoulders making it even more difficult to extract next time you try. In electronics we don't often come across really tight screws but even so it's worth being prepared.

There are two kinds of cross-point screws, those which accommodate a screwdriver with a pointed tip (Philips type) and those which demand a screwdriver with a blunt tip since the cross has a shallow bottom (the Posidriv and Supadriv types, which are *almost* the same). Trying to use the pointed screwdriver on the shallow head just means a chewed screwdriver. The other way about (blunt tool in a deep recess) is not so bad, but even so, the best option is to get the right tip.

Security screws are becoming more popular; if you chance across splined heads or 'special' screw heads then there's nothing for it but to stump up for the tool that fits.

Allen heads or 'hexagon heads' are very common. Unfortunately you may need to get both a metric set and an imperial set. Hex heads are difficult to chew up but don't be tempted to use a slack size, the risks of a rounded hole are great; a grim situation, particularly if it's a screw with a flush head, i.e. a 'set screw', since there's no chance of cutting a slot in it with a hacksaw if it is. If you get a rounded key on the other hand, cutting the end off to give you a fresh bit can get you out of trouble. In electronics you need the smaller sizes more often, the garage types are usually the larger ones. Save any keys which come with components (often a key is supplied with a knob or other item) as they will inevitably fit other things.

Screwdrivers bought as sets, in general, tend to be inferior to individually bought items. No, I don't know why either.

A really stubborn screw can be removed with an impact driver. Get a sensible one though. I once borrowed one which chewed my hand to bits because the flesh between finger and thumb got clipped each time the halves went together under the blow of the hammer. I didn't actually feel it until unsightly (although fortunately not permanent) damage was done. Motto: wear gloves. Unless you're building an industrial robot you won't want to put screws in that tightly yourself.

Incidentally, it's amazing how tight a really tiny screw can be. This is often the case with the tiny cross-points found in optics and I'm afraid you'll need a really decent set of jewellers screwdrivers to get those babies out.

Something with a hand-sized handle is a good start. Most jewellers screwdrivers are not worth it. I've even had cheap sets where the shaft rotated in the handle, if indeed you actually had the strength and agility to grip that tiny handle and turn at the same time. If you need tiny screwdrivers, pay sensible money for them and then moan like hell if they don't come up to scratch. An optician or a camera repair shop might put you onto a suitable make.

Angled screwdrivers are not needed a lot. Stumpy screwdrivers are good for access and easier to use; I find my stumpy crosspoint particularly useful for getting at the screws on the rear of computers without turning the things round on the desk.

A set of small spanners or a small adjustable is handy. For BNC connectors a very narrow spanner is needed to grab the connector body. You can get away with a wider spanner by forcing the sprung retainer away from the cable. Incidentally, it's amazing how quickly those spanners wear and they also splay (being so thin).

For an extra hand I use an old retort stand which fell off the back of a chemistry laboratory. The things with croc clips on are very light duty efforts. They wave about while you chase the components and are far too easily broken. Make your own less expensive ones from some good, stiff tinned copper wire and a wooden base if you really feel inclined. Also, the things which are supposed to stick to the bench by suction don't work too well. Use some brute mass instead.

I've never used a magnifying glass. I find that the distortions and reflections they introduce are worse than the smallness of the object in question. If you think you need a magnifying glass then try lighting the subject properly first; you'll probably find that works wonders. If you need one then the kind I would recommend is the 'jeweller's loupe' which fits over the eye like a monocle. They don't get in the way and you can switch to the other eye for ordinary viewing. Alternatively, the really large ones on a stand are all right. The 2" diameter ones are a fiddle to use as you need to position your head exactly to get the object into focus at the right place. Plastic lenses are easily scratched and once they get that way you may as well not use them; it will be worse than not having one.

It's always handy to have a couple of pairs of pliers; the snipe-nosed type are particularly useful, the square-nosed toughies with the serrated jaws coming a close second. Go easy on the round section needle-nosed pliers; they can crimp and flatten leads and almost cut them, even with moderate pressure. They're good for bending component leads to a nice rounded profile, however.

While on the subject of bending, you can get (or make) a jig which allows for consistent bending of component leads to a given pitch. If you do it a lot, you'll soon get the hang of where to bend using a pair of pliers. My advice is to bend at too great a pitch if anything; in this way, when you pull the component leads through the board, they will align themselves with no strain on the component body.

Rather than using the pliers themselves to bend the lead, which results in a big curve in the lead, use the pliers to grab the lead next to the component body and then push the lead over with a fingertip, close to the pliers, to get a really neat bend.

Finally, mole grips are useful for holding onto things or clamping them together and they need not be terrifically expensive.

11: The Mechanics of Electronics

The mechanics of electronics represents a special problem. First, packaging and all it implies is very often seen as subsidiary or unimportant when compared with the 'real electronics'. To some extent this view is justified, but a fine circuit, however sophisticated, will soon stop working if it is exposed unprotected to an outdoor or industrial environment of any kind.

The RAF seem to have an enlightened view of this; they have a specialized trade of 'electronic mechanic'. These people are just as important to the smooth functioning of the system as are the 'chip experts'. This is as it should be.

This section contains some useful information on mechanical assembly, enclosures, wiring, heat sinking and the design of printed circuit boards. The correct use of tools, as such, has been hinted at in Chapter 10.

Mechanical assembly

In this section I'll discuss all those mounting problems which seem to crop up, starting with the humble washer.

There are three reasons for using washers.

First, a plain washer will spread the tightening load across a larger area and also isolate the object being clamped from rotation of the fastener. This may be important to prevent some plastic bushes and the like from being crushed or smeared by the tightening of a nut or a bolt head against them.

Second, all kinds of washers can help prevent slackening of the fastening due to vibration, as mentioned in the previous chapter. A plain washer can be used in this context to back-up a spring (single coil) or star washer, isolating softer materials from them and preventing them from biting into and damaging plastic bushes and p.c.b.s. Metal surfaces can take direct contact with a coil or star washer. The actual spring and star washers themselves absorb any vibration.

Third, star washers can be used to cut through paint and oxidation layers to ensure good electrical contact between (typically) an earth tag and a metal chassis. Occasionally one comes across, in larger sizes, a nut with a captive star washer attached; quite useful.

Mounting a semiconductor package on a heatsink is a common assembly problem. Good contact between device and heatsink must be maintained for a low $R\Theta_{CH}$ (see below for calculations and explanation of terms).

Often, the situation arises where a heatsink must be placed on the outside of an enclosure but the device to be 'sunk' must be contained within. TO3-type packages have a distinct advantage here; their connections all lead through the heatsink to the inside of the enclosure, either of their own accord (the base and emitter connections, if we're talking about a power transistor) or via the clamping bolts (for the collector).

An insulating cap prevents any stray connection to the case of the TO3 device. Sockets with threads are also available for TO3 packages, making for easy replacement of the entire device from outside the equipment and with no soldering, yet another blow in the cause of maintainability.

The popular TO220, TO202, TO3P and other flat packages, however, are a bit of a nuisance in this respect; they need to be physically mounted inside the enclosure. Where this is the case, it is important not to attempt to clamp the heatsink to the outside of the case with the device mounting bolt. If done in this way, we're relying on the contact between the enclosure and the heatsink to transfer heat to the heatsink.

Apart from problems of gross flexing of the panels of the enclosure, especially where expanding due to heating, the heat sink and panel will both be microscopically rough and will only touch over a fraction of their area. (Silicone pads or greases and other heat conductive treatments are intended to be conformal over the intended area of contact and provide intimate contact by filling in these tiny rough spots.)

What is likely to happen, if we clamp the heatsink in the way that I have just described, is that the enclosure panel itself becomes the heatsink and only a little heat is transferred via the mounting bolt and the partial contact of the panel and heatsink, seriously compromising the thermal efficiency.

Instead of clamping the heatsink to the panel with the device mounting bolts, mount the heatsink over an aperture which has been cut in the panel and which allows the device to be fastened firmly to the heatsink itself.

There is another alternative; standing the heatsink off the side of the enclosure and hiding the devices behind it. This is not quite so good since there will inevitably be exposed wiring, or the possibility of exposed wiring, between the panel and the heatsink. The same applies to putting the devices on the outside of the heatsink and then cutting a hole in the heatsink for the leads or taking them over the edge of the heatsink. You need to be really careful about protecting such leads. Where the heatsink is inside an enclosure, of course, such precautions are not needed and wiring can be exposed within the confines of the enclosure.

The exact position of a component on a heatsink is not critical, although it is probably better off mounted near the centre of the heatsink so that heat can travel evenly away from it all around. When mounting several components on a heatsink, space them evenly, but give any components with heavier heat-sinking requirements a little more space to themselves.

Drill holes to clear the shank of the bush which comes with the device mounting kit, not a clearance hole for the mounting bolt itself. The bush keeps the bolt centred in the hole so that it does not touch the sides at any point.

Usually, it is important to electrically isolate the mounted component from the heatsink whilst maintaining good thermal contact. There are two situations, however, where this isolation effort might not be needed.

First, some components (e.g. high wattage resistors or 'isolated tab' TO202 packages) are already electrically isolated from their own cases and can be simply be bolted to the heatsink, with a smear of heat sink compound or a piece of silicone pad underneath. This is not for isolation purposes, note, but to promote more efficient heat transfer.

Second, it may be easier to isolate the heatsink from the chassis or the enclosure than to isolate the component from heatsink. This will be often be the case where the heatsink is quite a small one mounted on a p.c.b. or just pushed directly onto the component. As mentioned already, such a heatsink will probably not be earthed or isolated and must not be accessible to the standard casually-prying finger.

To save on using all those gooey heatsink compounds, you could use a silicone pad, either pre-formed or cut to the right size from a sheet.

When cutting silicone pads to shape, make them exactly the right size for the component, plus a little extra all round. Don't try to use a silicone pad which is smaller than the component. Where several devices are mounted on the same heatsink in close proximity, they could share a larger pad with several holes in it.

Tighten device-mounting bolts to just the right degree. A single hole mounting package may tilt slightly as the mounting bolt is tightened. If this happens, it will not give even pressure and therefore good thermal contact over the whole contact area. At the worst it could cause strains which will cause failure of some part, especially when heating causes differential expansion.

If cleanliness and freedom from metallic waste is important in the general run of things, it is triply important when mounting components on a heatsink. Metal chips, dust, cuttings or swarf, most often generated during the drilling or milling of the heatsink, can bridge across from the mounting stud to the side of the hole, thereby shorting the heatsink to the package.

You might just get away with a short from the mounting bolt to the heatsink if you're using a plastic package such as TO126 or the modern TO3P, where the mounting bolt might not touch any metal parts on the package, but don't bank on it. You certainly won't get away with it on TO220 or TO202 packages with metal tabs, or on transistors in TO3 or TO66 packages, where the collector is traditionally connected to the case, unless you really want to connect the collector to chassis?

A possible solution to the problem of shorts between the mounting bolt and the heatsink, if you're using higher voltages or are otherwise concerned about clearances, is to use a larger diameter mounting hole at the device mounting side. Drill part way through the heatsink from the component side.

When cutting such a larger diameter hole, be very careful not to cut all the way through the heatsink; a ruined heatsink may be the result. In fact, there is no need to cut beyond the point where the shank of the plastic bush protrudes into the hole. Also, be careful to drill from the component side of the heatsink. Finally, remember that removal of material by cutting the larger hole is bound to make the thermal resistance from case to heatsink RQ_{CH} greater, although it isn't possible to properly estimate by how much. For this reason, make the hole as small as possible and remove as little material as possible.

For a belt and braces approach, sleeve the mounting bolt, too, but remember that there may still be a gap between the bush and the sleeve the unless the sleeve is cut exactly to length. As a last resort, I suppose you could machine your own long plastic bushes which penetrate the full thickness of the heatsink.

The other problem with metal filings is that when the mounting bolts are tightened down, any metal caught between the component and the heatsink is likely to cut into the insulating pad. This is especially the case with the silicone rubber pads. While these are very convenient in use as they do not need greases, pastes and other preparations, they are far easier to cut up than the old-style mica washers. I've even seen the sharp edge of a TO3 can cut through a silicone pad - or perhaps I was a little too vigorous when tightening up. Any metal which cuts the pad will short the case of the component to the heatsink.

Making a front panel

The enthusiasm, or otherwise, with which your latest creation will be greeted, depends, rightly or wrongly, on what it looks like. In turn, the appearance of any equipment is almost always dominated by the appearance of the front panel.

First, as to design. The tried and tested way of allocating space to front panel components is to keep the displays near the top, any connectors or leads near the bottom and the controls sandwiched between them. A glance at the front panel of most instruments will give you an immediate impression of this general rule.

It's easy to see why things should be in this order. Cables must not dangle in front of controls or displays, so therefore connectors go at the bottom. In order to see the effects we are having on the system while we make adjustments to the controls, the displays must be above the controls. There may be exceptions to the rule but it usually holds.

Apart from this, the detailed siting of controls is dictated by their functions and their sizes. Group controls into related functions (engage in some more partitioning, in effect). Where there are complete sets of similar controls, such as those for the A and B channels on a scope, then make each group identical in form but clearly label them. Where the centres can be made to line up without causing other problems, then line them up (this makes the next stage, marking up, easier too).

If, in the course of this exercise, you get left with blank patches on the front panel, don't be too concerned; you're under no obligation to fill everything up completely. If you do end up with a sparse front panel, then leave more space around the groups of controls to differentiate them more strongly.

Avoid bunching everything into one corner or to one side, unless you've got a huge cabinet with just a few things to fit, in which case the best option is to put a modest group of controls at eye level and probably centred up in the horizontal plane.

Sketch the panel before committing to cutting the holes, to check for distances. Make sure that panel furniture is not going to foul the internal workings and that there is space for labelling. Labelling, incidentally, is usually done in a no-fuss 'sans serif' (no feet) fount such as Helvetica.

Marking up holes for the more complicated front panel components can be a menace, especially those which have an anti-rotation spigot which needs a separate hole drilling. To all those of you who can never get the smaller hole in the right place, I sympathize, although cutting a sensible hole with a flat in one side for the likes of a BNC is even worse.

The solution to both problems lies in precise measurement and in offering up the component in question to the panel as work proceeds. The guiding principle is: keep looking at what you're doing. As joiners say, measure it twice and cut it once.

Mark up the panel from the back, using a ruler for exact measurements. Every hole should be marked by the crossing of two well-defined lines, each parallel to an edge of the panel. Use a punch to make an indentation to hold the drill bit. A spring punch is a good idea, as it does not need to be struck with a hammer. In fact, hitting a spring punch with a hammer will wreck it.

To get the punch into exactly the right place, place a finger next to your marked cross-hair. Place the point of the punch across the finger and guide the point towards the cross-hair; you'll find that you can get really close with this trick. A similar thing arises with soldering, where the tip of the pinkie or the elbow placed on any steady surface steadies the hand sufficiently to get to grips with a fine piece of soldering. Just as an experiment, try placing the point of a punch, at arm's length, on a cross drawn on a piece of paper taped to the wall. How close do you get?

If you don't make such a punch mark, the drill bit will probably skate across the surface, unless you're using a pillar drill or drill stand, although even here it would not go amiss, as it is easier to see than the cross-hair of two scribed lines. Make sure the punch is perpendicular to the plane of the panel before striking it. If you're using an ordinary centre punch, a light but sharp tap with a small hammer is sufficient. It goes without saying that you must work on a flat, stable surface.

By the way, don't write, even with a pencil, on any panel which is to be on view. A pencil is sufficiently hard to leave noticeable indentations in the surface of the metal.

Before drilling, double-check to make sure that the marks you've made are in the right place. Hold the panel firmly. Better, hold it in a vice. Look out with larger sizes of drills; they may grab the panel, taking it out of your grip. Work with the panel horizontal if possible.

If you're using a pillar drill, set the drill turning before bringing it to the panel. If you're drilling by hand, place the stationary drill bit on the punch mark, check that the drill is perpendicular and then start up the drill.

For one-offs, you could use the rub-off transfers such as Letraset and similar. You need to have a perfectly clean panel to use transfers, otherwise they don't stick. Seal them to the panel with fixative clear lacquer after application.

Personally, I could never get transfers to appear straight. They always looked subtly crooked in some way.

So when I moved on to drawing panel markings by computer I thought that all my Christmases had come at once. Not only do I get straight text nowadays, but I can also have boxes with round corners, lines to connect controls, and other good things. Any of the small DTP or graphics presentation packages will do this kind of thing admirably; Corel Draw and Serif Page Plus are both to be recommended.

Once you've taken a deep breath and bought the equipment, a new front panel takes minutes to produce and costs little more than a sheet of paper. If you trim the edges off carefully and stick it down and then lacquer it or otherwise protect it, say by trapping it behind a perspex panel, it looks perfectly respectable. Laser printing is best; ink-jet printing is a close second, as it doesn't always have the same resolution as a laser, and the inks may sometimes smudge. Even a dense dot matrix version looks better than badly-done transfers. Depending on which kind of printer you have, you may be restricted to an A4 size.

Even if you have other ideas for the finished product, a paper mock-up is useful as a demonstration.

Another way of getting a sensible panel is to have your artwork made into a plaque. Usually, you see the coppery-looking ones with pictures of dogs or pheasants on them hanging in gift shops. However, there is an aluminium variety which looks neat as a front panel. Apparently, the apparatus used is a kind of photocopier. Again, there are size restrictions.

There are also a number of panel marking sheets on the market which, when exposed to ultraviolet light, turn opaque or transparent. You will need to prepare a transparency, however, and you will need a light box sufficiently large to take it. Some of these materials are fixed by chemical action, others by heating in water. You also need to have a bath of sufficient size.

If you're going to make lots of front panels, then think about getting them done outside. Some firms will do all the cutting and then silk screen the markings onto them. There are also options such as engraving, whether engraving of the front of a metal panel, engraving the rear of a perspex one or engraving plastic badges (Trafalite) which are then applied to the front.

These tend to be used on one-offs where cost is not a problem, where the panel markings are likely to be abraded, or on the really big stuff where other techniques don't have the reach. Apart from the Trafalite system, where an inner layer of coloured plastic is exposed, paint is rubbed into the cuts to bring up the markings.

Heat: getting rid of it

Issues related to actually mounting heatsinks and the workshop practices needed to ensure electrical and thermal integrity are given in an earlier section of this chapter. Here, we look at designing a heatsink into a system. There are two phases to designing in a heatsink. First, you must decide on how much heat sinking you need. Second, you need to choose a suitable heatsink and somehow accommodate that heatsink physically.

Does your circuit need a heatsink is the first question which must be answered. As we have seen previously, calculating power dissipation in any given component is quite simple, consisting of multiplying any currents flowing by the voltages which are causing them to flow.

Let's take as an example a voltage regulator pass transistor which is expected to carry 5A of current at the same time as it bears a collector-emitter voltage of 10V. By my reckoning, this makes 50W of power dissipation. There's probably about 50mW to 100mW of dissipation in the base-emitter circuit too, which I'm going to ignore as it's a lot smaller than the main power loss.

So what happens to all this heat? The heat is generated in the actual silicon die or chip itself. It heats the chip up and causes the temperature to rise; the temperature of the chip is called the 'junction temperature' (symbol: T_J). If the junction temperature rises above a certain limit (125°C for many devices) then the device will be destroyed.

Now heat tends to flow 'downhill' away from the semiconductor die towards regions of lower temperature. The die is bonded to the case, which takes some of the heat away, and the case is surrounded by air which convects heat away from the case. The trouble is, there are thermal resistances associated with each step of the way from the semiconductor junction, where the heat is generated, to the open air where the heat is finally disposed of.

Thermal resistance is expressed in units of °CW^{-1} and its symbol is $R\Theta$ (pronounced R Theta). Thermal resistance is analogous to electrical resistance; thermal resistances in series can be added together. Pressing this analogy a little further, temperature differential is the 'force' which drives heat to flow from one point to another and is therefore analogous to voltage differential. Heat itself can be thought of as analogous to electrical charge and heat flow as current.

Thus, adding thermal resistances:

$$R\Theta_{JA} = R\Theta_{JC} + R\Theta_{CA}$$

where the subscripts $_{JA}$, $_{JC}$, $_{CA}$ refer to junction-to-air, junction-to-case and case-to-air respectively. Sometimes the word 'ambient' is used instead of air; thus, $_{JA}$ would mean junction-to-ambient.

The heat generated must flow through these thermal resistances in order to escape. Unless the device junction is still heating up, we must assume equilibrium conditions. In fact we aren't really interested in how long the device takes to heat up; we usually assume that the conditions have prevailed long enough for the temperature to have stabilized. Under these conditions, the temperature difference T_{JA} is just enough to cause heat to be carried away at the same rate at which it is being generated.

We already know, from our example above, that our device must dissipate 50W of power; this much at least is dictated by the circuit we are using. I might think about using the popular 2N3055-type transistor; in this case, my Towers book tells me that this transistor has a T_{JMAX} of 200°C and a maximum power dissipation (P_{TOT}) of 115W *if I hold the transistor case temperature at* T_C *= 25°C*. This complicates the issue slightly; I would have liked a quoted figure for the $R\Theta_{JC}$ of the 2N3055, but I can easily work it out:

$$R\Theta_{JC} = (T_{JMAX} - T_C) / P_{TOT} = (200 - 25) / 115 = 1.5°CW^{-1}$$

Incidentally, this figure of 1.5°CW^{-1} is quite small; 4°CW^{-1} is more usual for TO3 packages generally. Typically, we then decide on a maximum permissible junction temperature and go on to calculate the maximum thermal resistance which will allow the heat to escape:

$$R\Theta_{JA} = (T_{JMAX} - T_A) / 50 = (200 - 25) / 50 = 3.5°CW^{-1}$$

I have assumed an ambient temperature of 25°C. So we are left to find a $R\Theta_{CA}$ of about (3.5 - 1.5) = 2°CW^{-1}. Typically, the case-to-ambient thermal resistance $R\Theta_{CA}$ for a TO3 package is 35°CW^{-1}.

A heatsink, then, is mandatory, if we are to reduce our $R\Theta_{CA}$ from 35 to a mere 2°CW^{-1}.

$R\Theta_{CH}$ and $R\Theta_{HA}$ represent the thermal resistances of the TO3 package case-to-heatsink and heatsink-to-ambient respectively. We want to minimize $R\Theta_{CH}$ of course. It can never be zero; using a variety of mounting methods, 0.2 to about 0.5°CW^{-1} seems to be the range of values we can expect. If we take the worst case figure of 0.5°CW^{-1} for this, then we are left with the need for a heatsink with $R\Theta_{HA}$ = 1.5°CW^{-1}. This is the figure quoted for heatsinks in the catalogues.

$R\Theta_{HA}$ is quoted for an ideal world where the heatsink fins are vertical and there is uninterrupted airflow around the heatsink (the heatsink is in 'free air'). Imagine your heatsink suspended from a very thin wire in the middle of a draught-proofed barn; this is free air. Unfortunately, enclosures are not like that and any judgements about how airflow is affected by the heatsink being near another surface are likely to be guesswork. For safety's sake, always assume that the airflow is lessened in real situations. Derate the heatsink to leave a safety margin; use forced air cooling with a fan blowing across the heatsink if you must.

One final point about choosing a heatsink: if you can, buy one with a central flat area, sufficiently wide to take the package you're trying to mount, rather than one with fins over the whole surface. Otherwise you are almost certainly faced with the job of milling or machining some of the fins flat to get the package mounting bolt(s) through the heatsink. You will invariably raise the heatsink's thermal resistance by doing this too.

One important point to note is that, if the power dissipation in a device exceeds the quoted maximum, then there is no possibility of ever getting rid of the heat, even if we had a heat sink of infinite sinking capacity. Where pulsed working or other circuit configurations are not possible, parallelling a number of devices to gain current or power handling capability may be the only option.

It's worthwhile exploring this idea of parallelling semiconductor devices in a little more detail. In days of yore, before MOSFETs were invented, bipolar transistors, when used in parallel, each had an emitter resistor of half an ohm or thereabouts to limit the emitter current.

The need for the limiting resistances is that, if one transistor has a gain slightly more than its fellows, then it will pass a little more current, heating to a slightly higher temperature. As the temperature rises so does the gain of the transistor. We enter a vicious circle of positive feedback; the transistor 'runs away', as the saying goes, taking more and more of the available current until it pops, either shorting out completely or going open circuit and throwing the load quite suddenly onto the others.

Thus, bipolar transistors are said to be 'current hogging' devices. MOSFETs, on the other hand, are 'current sharing' devices; their 'on resistance' rises with increasing temperature, tending to cut the current down. All things being equal, you may safely parallel MOSFETs with no current limiting resistors.

Sometimes, increased circuit complexity may result in a reduced heat-sinking requirement. Let's compare the temperatures of a single device mounted on a heatsink and a pair of similar devices, sharing the current, mounted on the same heatsink.

In the first case, the temperature of the single device is going to be:

$$T_J = T_A + R\Theta_{JA} \times P_{TOT} = 25 + 3.5 \times 50 = 200°C$$

Furthermore, the temperature of the heatsink will be:

$$T_H = T_A + R\Theta_{HA} \times P_{TOT} = 25 + 1.5 \times 50 = 100°C$$

Now the same amount of heat is flowing from the pair of devices in the second case, so the temperature of the heatsink is going to be the same in the second case. But the individual junction temperatures are each going to be:

$$T_J = T_H + R\Theta_{JH} \times P_{TOT} = 100 + 2 \times 25 = 150°C$$

Thus we have a junction temperature margin of 50°C now. The heatsink could, in theory, now have a $R\Theta_{HA}$ of 2.5°CW^{-1}:

$$T_H = T_A + R\Theta_{HA} \times P_{TOT} = 25 + 2.5 \times 50 = 150°C$$

Whether we use that margin to reduce the heat-sinking requirement as explained or whether we retain it as a safety margin will depend on the reliability criteria for our design.

Of course there are a number of different ways of manipulating the equations for thermal resistance and heat flow to get a result in the form you need. A basic understanding of heat flow and its electrical analogy and a facility with algebraic manipulation are all you need. Try a few examples of your own.

I've explored this in some detail since it's not something that is often mentioned. Thermal design seems to be very much an afterthought sometimes. Just try fitting a large heatsink onto a box which was not intended to take it! In fact, poor thermal design plagues many systems which are otherwise top-notch.

One of my favourite little tricks with linear voltage regulators is to use a 1A regulator where a 100mA regulator, purely from the point of view of current capability, would otherwise be adequate. The trouble with the weeny TO92 package is that its $R\Theta_{JA}$ is a rather large 200°CW^{-1}. If the ambient air temperature T_A is 20°C and the maximum allowable junction temperature T_{JMAX} is about 125°C, then we have about 100°C to play with and the maximum allowable dissipation is about 0.5W. If we're drawing 100mA from the regulator and have a voltage difference of 5V, that's our lot. Even at say, 80mA and 4V differential (320mW) we're crowding things a bit since the device is unlikely to be in free air.

Using a TO220-packaged device immediately drops $R\Theta_{JA}$ to 50°CW^{-1} and you don't need to use a heatsink. The more massive TO220 package itself provides the extra heat-sinking. A 0.5W dissipation, which would be borderline for a 78Lxx device, is fine for the 78xx 1A regulator, leaving a 1.5W margin. When you figure out that a 100mA 78Lxx series device costs as much as (if not more than) the 1A variety then it hardly makes sense to use the smaller unless you're very tight for space or need the fully insulated TO92 case.

True, the device is protected against excessive dissipation and will not be damaged under these conditions. But this is small comfort if the equipment you're building cuts out every few minutes as the temperature in the regulator builds up.

Incidentally, a p.c.b.-mounted TO92 packaged device will have different $R\Theta_{JA}$ with different lead lengths. Longer leads result in a higher $R\Theta_{JA}$, which seems a little odd, as you might reason that longer leads would push the head of the device further out into any available airflow. In fact, the decreased thermal resistance is due to closer proximity to the board and those lovely copper tracks which can carry the heat away more quickly via the shorter leads.

If you discover that your heat-sinking arrangements are going to be excessive, look at circuit alternatives which require less or no heat sinking. In particular, think about pulse operation in those systems which will take it. I'm thinking, in this context, about servo motor drives, switched mode power supplies and the like.

To move onto the actual siting of heatsinks. Fins should be upright where possible, to take maximum advantage of natural convection. The exception to this is where air is to be forced over the fins by a fan. Heatsinks are often on the outside of the enclosure, but need not be; where spot cooling of individual components is needed, heatsinks can be inside the enclosure.

A large cabinet can get rid of quite a bit of heat through its walls, and spot cooling of internally mounted power supply modules and the like is the norm. Be careful with internal heat sinking that the internal temperature of the enclosure does not rise to a point where further heat loss is compromised. Vents cut into the enclosure may be necessary to promote airflow.

Sometimes the heat-sinking capability of a plain metal enclosure or panel is adequate. Where heat-sinking requirements are modest, don't be afraid to do it this way. It's impossible to say precisely how effective a flat panel is at disposing of heat, but a good guess would be rather less than the worst heatsinks of similar size.

Very occasionally, you may need to provide a tiny heater in order to minimize temperature drift in a circuit. This technique is called 'ovening' and was often, in the past, accomplished by mounting a small light bulb in a small enclosure to provide a temperature controlled environment for the sensitive circuitry. Invariably, the temperature in this space was kept at a level higher than the expected maximum ambient. A much more reliable heating element is a resistor or other device, mounted on a heatsink if needs be.

I once had to build the circuitry to do this for a remote sensing infrared thermometer. The mechanics of the oven were built by a colleague and consisted of an old aluminium saucepan with a hole cut in the bottom for the camera to peer out of. The whole lot was foam insulated and a pulse width control circuit was mounted next to the heating element, which was a large resistor on a heatsink.

A word of caution: don't mount the controlling element in the ovened enclosure along with the heater, unless it is a pulsed or burst type. A linear control device will generate heat of its own, and often it will dissipate more power as the heating element itself demands less power, complicating the issue terribly.

Wiring

There are a number of topics which are usefully discussed under this heading: how much wire we need, how much space it will take, and what effect wires have on signals which are passing down them.

There is an easy way of estimating the amount of wire you need to use, which works for many situations.

Imagine that your wiring runs around the enclosure in straight lines which are parallel to the edges of the enclosure. If you have boards and components ready, mount them or use paper templates or otherwise mark their positions in 3D space. Alternatively, take measurements straight from your mechanical layout diagram.

If you now measure the distance from one point to another in all three directions and add them all together, that is the length of wire needed.

Allow a safety margin of at least a couple of inches on short runs, perhaps 10% on modest runs of a metre or so and a lower percentage on runs of more than a few metres (these will be uncommon).

This scheme works well until the route is not 'orthogonal' (always following one of three directions mutually at right angles) or the route doubles back on itself, in which case the doubled-back length is worth three times what the straight distance would have been (instead of just going there, you go there, back, and there again). Allow for this, as required.

The same principle applies on a larger scale with the wiring around buildings.

Recently I had to build a small avionics system which had upwards of a hundred wires in it. I reckoned that each of these was about a foot long. I asked my local supplier for 30m of hook-up wire, which looks like a heck of a lot on a reel. But it all went in the box; I ended up with a foot to spare! I wish I could estimate that well every time.

Wire takes space. I always reckon that a bundle of n wires, each wire of outside diameter $d$mm has a cross sectional area A of:

$$A = n \times 0.866d^2 \text{ mm}^2$$

which allows for each wire to take up a hexagonal space. This little equation assumes that the packing is perfect; in the event of any slackness in the loom, it will take up more space. Thirty cores of 7/0.2, each with an outside diameter of 1.2mm, will occupy 37.4112mm^2. Such a loom will just about fit through a 7mm diameter hole, with optimal packing.

If you're taking a loom through a grommet, allow for the inside diameter of the grommet, not the panel hole size.

Wire may also need to bend. A loom with thirty 7/0.2 cores in it is pretty stiff. Where I need to take a lid off something which has a bundle of wires going to it, I always try to fasten the loom firmly to each half. This prevents the strain of the stretching wires falling on such things as connectors and panel controls. Alternatively, a chunky connector, with good strain relief of its own, sited in one half may take the strain well and also allows complete disconnection.

I also leave a bight or bend in the loom sufficiently large to allow a good reach for separating the two halves of the enclosure and a radius for bending. Remember that when you close the lid this bend in the loom will need to go somewhere so you should allow clearance for it to lie without fouling boards and other items.

Where a wiring loom is being repeatedly bent, it may need to be guided in a chute or in a flexible conduit or tray to prevent it from kinking or getting trapped in any mechanisms. Allow a greater length of wire too, to allow a larger bending radius and less cyclic strain.

Wire also takes its toll on the signals which pass down it. The effects of cable resistance on d.c. power transmitted down a cable are described in Chapter 8.

There are actually two things which can happen to an a.c. signal travelling down a cable. First, the signal could be attenuated. Second, there may be reflections which cause ringing and standing waves to be set up. This latter is more important at higher frequencies and with digital electronics.

As well as the series resistance of a wire, there is also series inductance. There is also shunt capacitance, especially with co-axial cables. Values of shunt capacitance of $40pFm^{-1}$ to $400pFm^{-1}$ are common for co-ax, and series inductances of $0.25\mu Hm^{-1}$ are typical. Basically, the cable acts as a low-pass filter. At higher frequencies and longer lengths, you will need to account for this.

When you test systems which are to be connected by a lengthy cable, you can make a lumped-parameter simulator for the cable in question. Although rather crude, especially as they may radiate at higher frequencies, these can give you some insight into how the cable will affect your signals.

The equation for characteristic impedance is:

$$Z_0 = \sqrt{(L/C)}$$

from which it is a simple matter to calculate the series inductance per metre, given the characteristic impedance and the capacitance per metre. Test simulators can then be made using these values.

Reflections can be minimized by matching the impedances of the driving or receiving circuits, or both, to the characteristic impedance of the cable. Typically, we can use load matching or source matching. The pull-up resistors in floppy disk drives are a good example of load matching; they are used in conjunction with low-impedance (unmatched) open-collector drivers at the driver end.

Reflections are less easy to get rid of for the tracks of a p.c.b. In general, keep tracks which carry fast signals short. This will see you all right up to low MHz kinds of frequencies. If ringing is a real plague, try a series resistor in the track (a kind of source matching). This will attenuate the signal but it will also damp any ringing quite considerably. 27Ω seems to be a sensible value to try for TTL signals.

Bus systems are more complicated, since they are tapped along their lengths. When designing boards to be plugged into buses, keep the receivers and transmitters on the board as close to the bus connector as may be. Consider active termination, in which terminating resistors are connected to a threshold voltage, to minimize reflections on the backplane in bus systems.

There is an excellent discussion of these issues, and more, in a book by H. Stone, mentioned in Chapter 14.

When cutting wire to length, allow an inch or so for safety. I used to be very concerned about the waste of cutting off a spare inch and I tried hard to get it exactly right each time. Needless to say, I ended up wasting more wire (and time too!!) by having to discard the lengths that were too short.

This applies less to full-scale production where experience will have dictated an exact length by the time you've made a few articles.

Look at where the wire is going, and strip the insulation accordingly. Avoid having lengthy bare wires outside solder buckets and remember to include the insulation for most kinds of crimps. Experience will dictate how far to strip. If you're tinning the end of the wire prior to insertion in a solder bucket, then the insulation will shrink away from the heat, so it's often a good idea to strip a little less than you imagine.

Sometimes you need to preserve the overall braid screening on a cable. This is handy at the end where you're going to earth the braid or otherwise attach it to a shielding point. Depending on the length of braid available you can form a lead from the braid alone, or you can solder a length of equipment wire to it. (You should not necessarily attach both ends of such a shield; if it forms an earth loop, currents will flow in it, causing hum.)

To get the wires out from inside the braid, carefully poke a hole in the braid, close to where the outer insulation stops, without snapping any of the strands. Penetrate the inner plastic wrap, then lever out the handiest core, pulling it right through the hole. This is easier to do if the cores have been untwisted from each other. Use a small screwdriver or dental pick. The first core is usually the most difficult to withdraw. Once all the cores have been withdrawn, tug out the inner plastic liner, trim any filler cords off, pull any broken strands from the braid, stretch it neatly, and generally tidy up. Insulate the braid with sleeve if there's any chance of it touching where it's not wanted.

In some unscreened multi-core cables there's a handy filler cord which is often tough enough to use for the 'cheesecutter effect'. Strip back sufficient insulation to be able to grasp the cord with pair of pliers. Take hold of the end of the cable and pull this cord away from the end. With any luck, the cord will slice the insulation back for you. This works better in situations where the cable is quite long (since the cord will not slip out of the end of the cable easily) and where there is a foot or more of insulation to be taken off (it's hardly worth doing otherwise).

There is a common injury incurred in stripping outer insulation from cables, that of a stab or slash wound to the palm of the hand caused by trying to run a knife along a cable which is cupped in the hand. The blade invariably slips off, causing a perhaps minor but very painful incision. But you always lay the cable down on the bench to do that, don't you?

To insulate or not to insulate? Insulation, as well as preventing unwanted connection between conductors, can provide mechanical stability. For example, if your soldering is at all neat, then there is very little need to sleeve the pins on a Dee connector. However, heat shrink sleeving is a good plan in these places since it prevents the wires kinking and snapping, under strain or vibration, at the point where the solder stops. This is a vulnerable point in a soldered joint, especially with multi-stranded cores.

Finally, copper is quite soft stuff. In those confined situations where a wire stripper cannot be got in, small wires can be conveniently trimmed to length using a sharp craft knife.

The printed circuit

A complete circuit diagram is mandatory before embarking on a p.c.b. layout. So also are reasonably complete mechanical layout and interwiring diagrams. There are not many changes that can be made to a circuit in the light of the p.c.b. layout activity; there are rather more that could impact on the interwiring or physical layout diagrams. You may go round the loop once or twice before the big compromise between wiring, physical layout and actual p.c.b.s has been sorted out.

Associated with p.c.b. design are schematic capture and netlists. If you have got these features on any computerized system, then fine; I have restricted my comments here to what I call 'computer-assisted manual design', that is, the computer is merely a draughting aid and the poor human has to do all the actual decision-making. Computer assisted manual design is probably the cheapest and potentially most useful starting point as well as being the commonest p.c.b. layout method in use in smaller establishments.

Whether you are using a computer or not, there is a certain order of doing things which will save you time and frustration. I have coined the acronym OMILS to help remember this order. It stands for Outline; Mountings; Interconnect; Large components; Small components. Note that the first few things on this list are those things over which you may have little or no control. The last things on the list are the things over which you have complete control.

The board outline must be laid down first. If you like, use a thin track drawn right round the outside of the board. You need to draw this on all the board layers which are going to be used in the final product, including the component placement overlay if required. You will have to manually provide a board outline somehow, with 'no-go areas' where any mounting holes will eventually go, even if you have an otherwise fully automatic system.

Mounting positions need to be fixed next, even if there are no constraints on their position. For the usual, ordinary, small round holes use a large diameter pad to represent the position of the hole. Boards larger than about 160mm x 100mm (single Eurocard size) may benefit from more holes than the usual 'hole-at-each-corner' arrangement.

Try to position mounting holes close to the proposed positions of any board-mounted connectors and any particularly heavy components, so that these exert less leverage on the board.

Distinct stages of the design process are influencing each other, yet again, here. You may go round the loop once or twice before compromising between mounting holes, the positions of larger components and your ability to squeeze tracks past other things.

Look up the data sheets when using nylon pillars with latching or snap-in tops, or take up a sample and try it on for size in a scrap p.c.b. Some of these pillars need surprisingly large holes to work properly.

Metal pillars need good clearance to nearby tracks if they are not to be accidentally connected to those tracks. Occasionally I've seen deliberately metallized areas around mounting holes, the intention being either that a connection is made down the pillar to the chassis or enclosure, or to earth the pillar to prevent it from re-radiating any r.f. Doing this for the former reason is unreliable; don't be lazy, use a wire instead. The added advantage of using a separate wire is that it gives you a choice, at a late stage, of whether or not to actually bind the board to the chassis.

Connectors need to be placed next. There may be some constraints on connector positioning; after all, they need to be easy to reach, otherwise there's no point in having them. Even when you have a complete netlist, you may need to place some of the components manually, especially connectors, which are most often along the edges of a board.

Leaving insufficient space for connectors is a common mistake. Occasionally one comes across a connector of which the board mounting part is physically smaller than the other part. The mistake is to fit the connector too close to some other nearby component to allow the mating half to get in, so do allow for that.

Closely-spaced connectors can suffer from inaccessibility to fingers too. Try your own fingers in and see if they can grip the part effectively. Remember to allow for any overhang of the non-board-mounting half of the connector if it sticks out from the board at all. This is, properly speaking, more a decision that has to be made when drawing up the physical layout diagrams. Is there space to withdraw the connector with the board *in situ*, or does the board need to be removed from the enclosure before the connectors will come off?

In a completely manual situation, sketch out your design on a piece of scrap paper. Do it full size or twice size and use differently coloured pens, or pen and pencil, to signify different layers of the board if needed. A simple sketch may save you an awful lot of fiddling, as it allows you to see how many tracks are needing to squeezed past any given point. Rip-up-and-retry and rubber-banding are wonderful tools when you have a computer doing the placement for you but is one of the most thankless tasks ever to do manually. A sketch can be useful, though, even in the computer-assisted situation as a starting point to assist placement.

In computer-assisted manual situations, place components with no tracks first. It's almost inevitable that you'll have to undo some of them before you've finished.

On a computer you can afford to make up subsystems and then juggle them. For instance, you may have the layout for a nice filter completed, but you're not sure how it's going to fit with all the other things on the board. Pick up the whole lot and move it outside the board boundary while you work on other things, moving it back into the board afterwards, rotating it if needs be. Better yet, use a copy, preserving the original layout; you can delete the scratch-pad areas after you're satisfied with the result.

Occasionally you're in the position of knowing that you'll need more than one board, but you're wanting to balance the density of the boards and you're not yet convinced how much you can reasonably fit on any one. Shuffling subsystems from one board to another is quite in order here, but make sure that changes are reflected in your interwiring diagram, which will inevitably change to suit.

Repeating motifs are easy to accomplish by computer. A bank of filters only demands that you lay out the first one; the remainder can be copies of the first, with power supply tracks and such popped in afterwards.

A board outline which is in common use can be stored on disk to be used as the starting point for further designs, in the knowledge that it has been tried and tested previously, i.e. it really fits!

Remember the 80-20 rule? Well it works for p.c.b.s too, with a vengeance, as far as manual and computer-assisted manual working go. Getting those last few tracks pushed through is a bore. You have to live with that, though, and a good sketch at the outset can assist.

Print out a draft copy and make sure you've not missed anything. It's also a good idea to get a draft print-out when you're half way there, sketching in the remainder of the tracks and taking note of any odd things which need to be sorted out.

Unless you can really rely on what the computer tells you are component footprints, plonk components down on the actual print-out to check spacings.

Laser printing onto transparencies can be a bit of a liability, as the transparency may not take the laser toner very well; some pinholing may be the result. If possible, print on a piece of good quality paper and then transfer photographically (using lithographic film, probably) remembering that the finished transparency must be *exactly* the same size as the original.

If you insist on printing direct to transparencies, then you should get the heat-resisting type which are suited to photocopying. An ordinary transparency melted into the works of your nice laser printer represents a lot of expensive damage.

12: Construction Methods and Prototyping

The construction methods outlined in this chapter are typically those used for prototyping and testing, although they may be extended to small production runs if needs be. First, however, we will look at the prototype itself, examine it uses and nature, and try to derive some rules about the construction method most appropriate.

The decision about how many of a given article are to be produced reflects strongly upon the choice of construction method, in that some ways of building a system demand more up front in terms of investment in the design process and in tooling up. These rather commercial constraints have as much influence on the choice of construction method as technical constraints do.

There may be many prototypes in existence, each highlighting some aspect of the finished system. However, unless the final or 'pre-production' prototype is built using the same construction methods as the final product, it is not a true test of the functioning, or of the production requirements, of the final product.

What is a prototype?

There may be occasions when the prototype is going to be the only article produced. In these cases, you would tend to minimize the investment in design, since this would only be recoverable over this single article. Tooling up, unless those tools are to be used on other projects, will tend to be minimized in this situation too.

The exception to this business of minimizing is where the design itself is the product; I'm thinking here of magazine articles and similar situations where the prototype is intended to test the viability of a design which *someone else* is going to make. In that case you're being paid for the design work, not the actual device, so the design itself takes on a special significance.

Having disposed of those rather special cases, let's go on to examine what a prototype is, in the sense of 'first-of-a-kind' rather than 'one-of-a-kind', and what its relationship is to the finished article.

A prototype is intended to test something. It may test the access to the interior of the enclosure, the actual working of the circuit, the ease of use of the system, the robustness of the mechanics or indeed any physical feature which needs to be verified. It may also be use to test the reactions of potential customers.

A prototype can range from something which looks most unlike the finished article to something which is, to all intents and purposes, indistinguishable from it.

If just testing circuit operation, a prototype may not be encased in its proper enclosure and indeed may bear no physical resemblance to a finished system. This 'engineering prototype' is the usual state of student projects. There is nothing inherently wrong with this; after all, a student project tests the ability to approach a problem in a given way. However, a typical but unfortunate feature of a project at such a stage is that it only works when the inventor is present, a strong argument for the existence of psycho-kinetics, perhaps.

Where it is intended that a properly working system be produced, then a modest amount of extra effort at the outset and a little extra care in testing and construction will reap enormous benefits. A real working device is by no means the natural outcome of a rigorous academic investigation. Anyhow, I digress.

At the other extreme, a prototype may look and feel just like an example from the production line and exists as a demonstration of technical fitness and performance to senior management, supervisors, colleagues in charge of production and potential customers. Such a pre-production prototype is, in all but name, the first example of the finished article. Note well, the success of the prototype says nothing about commercial viability.

For a one-off, the prototype often *is* the finished product, with the proviso that changes might be made to it and embodied in any further items which happen to be produced. A prototype is eligible for change. Whatever it looks like, the prototype must work sufficiently well to test whatever aspect of the design is called for at that stage.

A prototype is complete in all the respects which are being tested at the time. Fragments of a circuit on a breadboard do not really constitute a prototype.

Some systems are so large that changes are inevitable from one individual to the next. I doubt very much whether there have ever been any two space shuttles, for example, which are identical to each other. Perhaps this is a rather extreme example, but it illustrates the point; larger systems suffer more variation from the prototype as more is learned about the way they work in practice, as the possibility of making incremental improvements are discovered and as these are implemented.

Beware of making incremental improvements to small articles. Where would Sony be today if every Walkman had a slight change made to each one to make some minor improvement? Entire *batches* may be slightly different, but not individuals. So think about the scale of the job, too.

For this reason, most personal computers rolling off the production line are absolutely identical, even to the point of all the screws being tightened to the same torque. Customization, where it exists, consists of merely plugging in options, which are themselves volume products, and of re-positioning a few internal links or switches. The design is to some extent deferred, the end user making certain choices about the final nature of the product.

It's worthwhile at this point to take a look at the relationship between system complexity, production quantity and the likelihood of individuation. Basically, as more human effort is invested in any individual item, so that item is more prone to suffer improvement or individuation which differentiates it from its fellows.

This is either a problem or a benefit, as explained above, but the important point is that any variations must be documented. The most efficient way of doing this is to issue a standard set of documents along with appendices noting the changes, rather than making up a complete new set each time. CAD and computing generally make it possible to make changes quickly, but be careful not to overwrite the standard document set with a variant. On larger systems, with many users each making changes, it is the system manager's responsibility to disallow writing to the standard document, unless specifically asked to do so.

So we may not get beyond a prototype. This may be because we intended the prototype to be the finished item, in which case it must be a complete working unit even though changes may be wrought. Or because during prototyping it became apparent that production was not viable (the prototype was part of a feasibility study, perhaps). In either case, documentation must be up to scratch, even if only to present a case for abandonment.

Construction methods

The smaller electronic components may be most conveniently mounted on some kind of sheet material such as stripboard, prototyping board or a purpose-designed printed circuit board (p.c.b.). Where a number of these boards exist within the same cabinet they may be stacked somehow or slotted into individual slots in a 'card frame'. For convenience I have classed all ready-drilled stripboards, prototyping boards and wire wrapping arrangements as prototyping boards since changes are more or less readily accommodated during construction. Like the microprocessor, they are 'deferred designs', intended to provide a basic functionality which is customized in use. Plug-in breadboards are not prototyping boards since they lack even the potential for permanency.

Some decision needs to be made about the nature of the board or card used in construction. Such a decision will be based upon likely production quantities (is it a one-off?), the circuit technologies in use (are there many i.c.s?), what the robustness or reliability requirements are (will stripboard break under these conditions?), whether or not you are prototyping or approaching the finished article and what your resources amount to (have I the time and the equipment to go for a p.c.b.?).

A p.c.b. loads the start of the development cycle, in that there is considerable design effort in a p.c.b. before any construction proceeds. Other systems of mounting put the load onto later parts of cycle, i.e. the preparation may be minimal and the design may grow organically and spread across the board as work progresses, as in the manner of wire wrapping or stripboard.

A block diagram should be prepared before starting. Except in the case of the p.c.b., where a complete and tested circuit is mandatory before embarking on design, a circuit diagram could evolve as the circuit is built. Even so, the circuit diagram should be roughed out before starting construction and it must be kept up to date as construction progresses. Some attempt should also be made to estimate the amount of board area needed before starting, otherwise physical partitioning of the system operates at random, with undesirable consequences for keeping track of the work, trouble-shooting and possible replication.

Unless there is a breadboarding phase, however, be prepared to rip up those portions of the circuit which cannot be made to work and start over. Experience will dictate those portions of a circuit which are novel (or novel to the designer, at least) and which therefore merit closer attention before committing to the soldering stage.

The risk with a p.c.b. of is one of over-commitment and the possibility of changes being expensive to re-tool. Where circuits have been designed to be non-critical, some changes can be wrought simply by altering component values. Partitioning helps considerably here by isolating circuit functions and minimizing interaction between individual circuit parts, thereby making the need for 'catastrophic redesign' less likely.

Stripboard, wire wrap and their derivative prototyping systems allow changes to be made during construction itself. They also increase the risk of variations or outright mistakes cropping up during construction, where several 'identical' articles are being made.

Using standard sizes of prototyping boards, which slot into card cages, can alleviate mounting problems. Often, where a prototyping board is intended for use in a standard cage it will have a pattern laid out for the cage's standard connector rather than all the space being devoted to the matrix of holes. Such boards are available, for instance, to plug into most standard computer buses.

Invariably, flying leads are the interconnection method of choice for stripboards; the number of board-mounting connectors which can be applied successfully to stripboard are few indeed. Besides, stripboard's fragility may prevent the mounting of any really heavy components or components to which a degree of operating force might be applied, such as connectors.

P.c.b.s on the other hand allow the finest possible control over mounting options, card sizes and special features and are mandatory even at the prototype stage if things like ground planes, guard rings and the like are needed.

Even where the electronics in a finished article will be committed to p.c.b., a prototype may be constructed mounting the circuitry on stripboard. However, this stripboard version is unlikely to resemble a finished p.c.b. physically and is really only of use in verifying the circuit operation. Otherwise, the circuit or fragments of the circuit will need to be proven on the breadboard, enabling a direct commitment to the p.c.b. with minimal likelihood of large changes needing to be made. The only problem with this transfer being the capacitances within the breadboard itself which will be absent from the p.c.b.

Stripboard

Stripboard, as the name suggests, carries a number of conductive copper strips on its non-component side. These strips are usually arranged on 0.1" centres and are perforated on 0.1" centres too, resulting in a 0.1" matrix of holes, which is convenient spacing for d.i.l. packages and any other components with 0.1" pitch pins whose rows are not too closely spaced. The strips run the length of a complete piece of board rather than the width. It can be obtained with a 0.15" matrix spacing too, which is more useful for larger components.

Component leads are passed through the board and soldered and cropped in the usual way. Where continuity to a neighbouring component is not needed, the strip is interrupted using a track cutter which is essentially a twist drill; indeed, an M4 drill or similar makes quite a good track cutter if the genuine article is not to hand. But watch your fingers on the drill flutes; for preference, hold the drill in a spare chuck.

Usually with stripboard, quite a good proportion of the board area and the holes are taken up by connections running across the board. These are usually made up from insulated 1/0.6 solid core wire, although shorter links, which do not represent such a shorting hazard as the larger links, are often made from ordinary tinned copper. In fact, cropped component legs can be usefully pressed into service in this role.

Stripboard with copper strips running across the board on the other face is available, although it is not common. It has the advantage that connections which need to run across the board can be made just by shorting through the board with a pin. It has the disadvantage that tracks need to be interrupted on both faces and there is greater chance of an inappropriate connection being made.

Each strip can carry up to about 5A before overheating occurs; personally, I would rather that currents of this order be carried in separate wiring where this could be arranged, even if only to lessen voltage drops and possible interference due to power supply coupling.

Don't get the idea that using stripboard is any easier, in the long run, than designing a p.c.b. A circuit mounted on stripboard is every bit as difficult to get right as the design of a p.c.b. *of equivalent complexity.*

Alas, there is still some preparation to be done before committing components to a humble stripboard. You need to have at least a rough circuit diagram. Convert this to a component overlay, showing the sizes and positions of the components (the component's 'footprint') and marking the holes where the leads or pins go. Also, mark in the positions where the strips will be cut. A particularly common mistake with stripboard is to forget to cut strips where needed; proper planning helps to prevent this, and helps also to locate mistakes quickly where they have crept in.

Make up a few sheets of the overlay. Do them in pen and then work in pencil so that you can correct easily. Making them twice life-size makes it easier to see what you're doing. Use a code to show those holes occupied by components and those positions at which the strip is broken; I fill in the occupied holes and use a cross for interruptions.

Before placing any component on the sheet, mark in the positions of any mounting holes, unless the mountings are not critical, in which case they can lie wherever there is a convenient space. Do, however, space them as evenly as possible and don't forget about them altogether.

Also, reserve one complete strip for each power supply rail; you'll find this very convenient on larger stripboard projects as you'll just need to drop wires to the i.c. pins to provide power. I use a dot from a coloured pen to remind me which of the rows is positive, negative, etc. If you keep the rails adjacent to each other you'll also find it very convenient to solder in decoupling capacitors. Where there are many power connections to a few rails, say on a large digital system, it may be best to have two pairs of rails and build the circuit in the form of two segregated upper and lower halves.

The golden rule is: connections running the length of the board are provided by the strips, connections across the board (i.e. from one strip to another) are provided by wire links.

I.c.s are always placed with their long axes across the run of the strips and there are almost always interruptions between every pair of pins. Discrete components usually run in the same general direction as i.c.s. but can be usefully placed along the strips on occasion. You must, of course, interrupt the strip between the component leads - otherwise the component will be shorted out! Placing components diagonally is unnecessary in a well-planned board and is untidy as well.

Don't forget to place any wires running across the strips, these occupy space, too. If you have some computer drawing program which can be set up with your own symbols and you think that, due to the size of the job(s), it warrants the time spent setting up, then by all means use it to lay out your stripboard designs.

Strips are cut by pressing the point of the cutter into an existing hole and twisting. As with filing, you're cutting not shuffling, so a couple of sensible twists might do. Go steady until you get some practice; stripboard is soft stuff. Check your efforts after cutting to make sure that all the copper is removed and that there is not a thin whisker of copper remaining at the edge of the cut to provide an unwanted connection. There is no need to cut all the way through the board (which weakens the board further) although stripboard is so soft that this is easy to do if you want a mounting hole in that position.

A track is always interrupted at a hole and occupies a whole 0.1", the only exception to this being occasioned by mistakes made during incomplete layout planning. In that situation a craft knife can be used to scrape an emergency interruption between two adjacent holes!

Cutting strips before soldering is the best policy, as it is difficult to cut a strip close to a soldered joint without disturbing it.

Stripboard is fragile. As well as being made from s.r.b.p. (resin bonded plastic) which is not as strong as the glassfibre now used in all p.c.b.s, the multiple perforations weaken it further. This fragility can be used to advantage when cutting stripboard. Score on the copper side, across the copper tracks is best, and snap over the edge of the bench. There's no need to get all excited with a hacksaw for stuff as weak as this.

One board-type akin to stripboard, variously called 'development boards' or 'proto-boards' - has interruptions at predefined i.c.-type spacings. Each pin of an i.c. has a small strip with three holes assigned to it: one for the pin of the i.c. itself; one for the wire arriving; and one to chain a wire going out. In this way it's always possible to connect together as many i.c. pins as needed. Passive components have to make use of spare i.c. sections.

Such a board is ideal for 0.3" spaced logic i.c.s, although fitting a 0.6" pitch i.c. is a problem. Often, boards such as these have complete power tracks for conveniently picking off zero volt and +5V supplies and, where they are of a standard rack size, they may have a pattern laid down for the standard bus connector.

To summarize: you've probably decided to use stripboard for one of the following reasons: it's cheap, certainly in terms of hard cash if not in terms of time; you're only making one item; a professional *looking* job is not so important (the performance may be brilliant and the actual circuit design itself may be a work of genius); you're concerned about the possibility of changes needing to be made; you want to see results you can use right away and are not prepared to invest time at the outset; you probably do not need to plug into a bus and are content, in the main, with flying leads as interconnections; you did a bad thing and just succumbed to the temptation to git solderin'.

Wire wrap

Wire wrap is simply the art of wrapping wires around the legs ('spills') of i.c. sockets to provide the necessary connections. Although that might not sound very secure, in fact a properly done wrap is a nice tight connection and very reliable. Although suited to hand crafting, automated wrapping systems have been known.

Wire wrap used to be used for full-scale production, although its labour-intensive nature makes it better suited to small runs and prototyping. You'll come across it mainly in the older medical, scientific and oil-field gear which was expensive, produced in limited quantities and much subject to change and modification.

If you aren't sure whether or not the system is going to evolve through a number of field trials, then wire wrap is even easier to change than stripboard. It also tends to be a little more robust.

Wire wrap is a great prototype method where there are loads of i.c.s, since special wire wrap sockets with long 'wire wrap spills' are easily obtained, as are dual row header plugs (mating to ribbon cable IDC sockets) with similar spills. Wire wrap is far superior and quicker to use than stripboard in the sense that changes are very easily made; in effect, it needs even less pre-planning than stripboard. But there will be wires everywhere, so if you don't like Italian food, forget it.

Passive components or indeed any components not in a regular d.i.l. or s.i.l. package cause obvious problems for wire wrap. One way around this is to use a plug-in carrier whose upper surface bears cup-shaped connections reminiscent of the rowlocks on a rowing boat. These carriers convert the individual components to a d.i.l. format. Fine, if your passives will lie nicely on a 0.1" spacing and are no longer than the width of an i.c.; otherwise, you waste a few pairs of contacts or may even occupy a whole carrier just providing for one fat capacitor.

Another disadvantage of this carrier method is that it tends to concentrate passives which would otherwise be distributed around, and close to, the circuits to which they properly belong. Don't put decoupling capacitors on such a carrier! In all fairness, decoupling capacitors can be connected across rails which are provided for the purpose around the edges of any board designed for wire wrap.

You can make a kind of hybrid stripboard/wire wrap system if you want by using wire wrap sockets for i.c.s but soldering passive components direct to the board. Tack the wires to the ends of the passive components underneath the board. This can be particularly advantageous where there is an easy division between a predominantly digital section of a circuit and a neighbouring analogue system where passive components are used extensively.

A hand tool or motorized pistol grip wrapping tool is mandatory for wrapping; you cannot wrap without something like this. The motorized tools are battery powered and the more expensive types are rechargeable.

I remember that my first experiences of using an automatic wrap pistol were a disaster. I never worked out why; but the second time, with a different tool, was fine. Although touted as 'automatic', motorized tools are hardly that; actually it doesn't speed you up much to have a fancy wrapping pistol unless it also cuts and strips the wire. If you ever want to unwrap (and there will be times!) you'll still need a hand tool for that.

For all the wrapping I do (boards with a dozen or two dozen chips perhaps), a combination wrap/unwrap hand tool, with one end for wrapping and one end for undoing wraps, is all I need. There is a wire stripper of precisely the right dimensions in the handle. There are some combined cut and strip tools which guarantee to strip the right length of wire for the wrap; I just use the end of the tool itself as a guide to stripping length.

You need to do some preparation for wrapping. I.c. sockets should be mounted before wrapping starts. These can be on a really tight spacing if you're strapped for space. I leave uncommitted sockets dotted about my boards ready to be pressed into service for whatever purpose suits at the time. I only solder two diagonally opposite corner pins to retain i.c. sockets, just in case I want to remove them at any time. Header plugs are a different matter; if the end result is going into service rather than staying on the bench as a test-bed, you may decide that the extra mechanical support gained from a complete soldering job is worthwhile.

To wrap, first cut the wire to length by laying it out along its intended route and allowing two inches for wraps. I tend to follow neat routes at right angles rather than direct diagonal point to point; besides looking neater, the wiring fits better if channelled between i.c.s and you can find your way around better. Allow for some slack too; there's nothing quite as frustrating as being just short of the mark after wrapping one end of a wire down.

Don't be tempted to stretch wires around corners or wrap using the bitter end of the wire, which leaves naked fragments of wiring sticking out of the wrap turns. I once came across a situation where the current going into one end of a wire did not equal the current emerging from the other. Since this flouts all the laws of physics, I was a little mystified. Mr. Kirchoff would not be pleased, I thought.

Rather than admitting to the poltergeist theory of electronics, I investigated further; lo and behold, the wire in question took a tight turn around another spill, whose corner had sawn through the insulation due to vibration and shorted to the internal conductor. So as I say, don't wrap too tight.

Strip both ends of the wire. An inch is plenty. Poke the wire into the smaller, offset hole in the wrap end of the tool. Bend the wire neatly away from the end of the tool and then push the tool right down to the board or to the top of the previous wrap (three or four wraps can be put onto any one spill).

To prevent the wire from just being carried around the spill rather than wrapping, hang onto the loose end by trapping it under your thumb or otherwise grasping it as you grip the board. Turn the tool clockwise to wrap. Be careful; you can actually successfully wrap counter-clockwise with most hand tools, but then the unwrapping tool will not be able to pick up the end of the wrap to loosen it! There is no need to apply excessive downward pressure to the tool, but apply a gentle weight to keep the turns touching as the wrap is made.

Most tools nowadays produce a 'modified wrap'. This means that the first turn or two around the spill is of insulated wire, the remaining turns forming the connection. This results in better vibration proofness (not perhaps very important in a prototype). The pressure of the wire on the corners of the spill breaks any oxide coating on the wire and gives a 'gas-tight' connection between the two.

I like to wrap all my 'computery' circuits instead of breadboarding them on the usual plug-in breadboard, for two reasons. First, the number of wires in a such a circuit, rigged in traditional breadboard fashion, would be fearsome to contemplate and very difficult to work on (although I did once produce my own floppy disk controller that way, back in the days when disk controller chips were newly announced and incredibly priced). Second, I prefer to use a board which plugs directly into the bus of the computer I now use, which saves struggling to get the bus connections out to a breadboard.

I also tend to leave fragments of circuits which I know I'll use again (address decoding, etc.) ready-wrapped on the board to save time; after all, why should I repeat myself at length? Those bus connections which do not change can have wiring soldered directly to the pads and wrapped to the appropriate pin of the decoding circuits. Those which may be re-routed for different purposes are provided with a wire wrap spill, as are all of the supply connections.

Although I chain signals, I tend not to chain power supplies. Two or three chips are all I like to have on one power supply connection; in an ideal world I would take a separate wire from an appropriate wire wrap post to each chip. This I would particularly like to do for the zero volt line on analogue systems; see my comments on ideal power supply wiring in Chapter 8.

A variation on the theme of wire wrap uses solder posts. Point-to-point wiring is put in place and then soldered, melting the insulation at the joint. This is a rather permanent affair and not easy to get undone once it's made. Note, you do not need to solder an ordinary wire wrap joint.

One commercial automated wiring system does similar things with wire, the result of which is very similar to a p.c.b. in appearance. The advantage is that the wiring is arbitrary; you can connect anything to anything regardless of routeing considerations, number of layers, etc.

To summarize: you have probably chosen wire wrap because you have a dense digital circuit with few passive components; you do not mind a little expense at the outset; you are concerned about making changes easily; you are not going to make many of these boards; you have space for the long spills underneath the board or between the boards; you are not working at frequencies of more than a few MHz.

Printed circuits

There is nothing quite as good as a custom-made p.c.b. for mounting electronic components, which explains its widespread use, especially for equipment produced in quantity. While working prototypes might be made up in a number of ways, you will probably still want to use a p.c.b. for the finished article.

The actual mechanics of good p.c.b. design are discussed in Chapter 11. Refer to this for more information on how to actually produce good p.c.b. design.

You have probably chosen to use a p.c.b. for these reasons: you are going into production and want to minimize the build effort for every item; you need the robustness of p.c.b. construction; you are still prototyping but it's important to see what the finished product will be like in every detail, i.e. you need a pre-production prototype; you do not mind initial expense and doing a lot of work before grabbing the soldering iron; you are using a circuit whose correctness you are reasonably certain of, or which you have proven on the breadboard; you are working in a cramped space and can't afford the space overhead of other methods; you need the very best control over component spacing and footprint and do not want to be constrained to the rigorous 0.1" matrix imposed by stripboard and its derivatives; there are features of the circuit (striplines, guard rings, shunts, planes, etc.) which cannot be accommodated on a stripboard or with wire wrap; you just want to make a neat job, thanks all the same.

Testing as you go

If you're committing to p.c.b., a lot of the testing will have been done at the breadboard or first prototype stage. If you are using wire wrapping or stripboard, then there is a good case for testing as you go. First, errors may more easily creep in with these latter; second, it is easier to test partly completed systems. Insufficient testing can be the ruination of an otherwise good project.

If your partitioning is good then each of the sections of your circuit should have an obvious division. Build one section at a time and then test to see that the inputs relate to the outputs in the way that you require.

If necessary, uncouple sections from one another (remove the component or wire link which joins the output of one to the input of the next) and inject simulated input signals. This will give you a reliable and complete test where real input signals are not readily available or where they do not give the range of inputs over which you want to test.

If you think you're going to be soldering and de-soldering a lot, use a two-pole header and a shorting link to switch between an injected input and the output of the previous stage. Document these links so you don't get confused!

By the way, you can occasionally justify knocking up a little test box which either generates signals itself or just provides a neatly labelled interconnection scheme converting from one set of plugs to another which makes testing easier. Mostly, these test boxes need not be complicated; providing your customer with one is an excellent public relations exercise!

Don't progress to building the next section of a circuit without sorting out current problems, especially where sections are chained in a series of processes, each of which depends on inputs from a previous section.

Physical segmentation may not apply; for instance a quad op-amp may be logically segmented to have one device devoted to an amplifier, one to a filter system, one to an absolute value ('perfect rectifier') circuit and one to a comparator circuit. Apart from the fact of a common supply, these four op-amps can be treated as individual entities as far as segmentation and testing is concerned.

Making changes

Breadboarding or testing as you go (either as a prototype or as a finished article) always involves changes, unless you're very familiar with the circuit in question and have been very thorough in your circuit calculations. Always note any changes as you go, either noting them directly to your rough diagram or transferring them *en bloc* to the appropriate places whenever more than a few have accumulated.

First documentation looks rough and it will be in note form. Don't be afraid to scribble on it, but make the scribble legible. Documents intended for user guides and for communication need to be a little better presented.

There is a pressing need to modify some equipment, particularly experimental apparatus which various people will use for things it wasn't intended for. Some gear might have suffered (literally) modification at the hands of whole generations of experimenters and in this case the documentation may bear little resemblance to the actual circuit. Couple this with the loss of some diagrams and the insertion of one or two which relate to some cousin or distant ancestor of the gear in question and you have a ripe recipe for colossal confusion.

Recognise the value of up-to-date information. Lodge copies of diagrams with the customer or user, especially in the case of one-offs. Keep a personal set which you can modify at will and then see that any other copies in existence match.

13: Going into Production

So we've got a watertight design; the prototype is made and looks good and functions perfectly. The product is easy to build and to maintain. The documentation is up to date and we have a full set of guides and manuals. Technically, we've got it taped. What's next?

There are a number of things to ponder upon. One is the actual making of the article itself (are we competent, can we finance it, have we the tools and so on). Another is how we set about employing people or taking on partners and how we look after them and provide a safe, sociable and interesting work environment. The other is the selling and distribution of the product, which has as much to do with the environment in which the organization operates as it has to do with what goes on behind the factory doors.

So will the world beat a path to our door, and if they do, are we in a position to fulfil the demand? This chapter addresses some of these concerns.

Will it fly? Feasibility studies

One could argue that the feasibility study is the first thing on the list for any project. Feasibility studies, however, involve some of the stages of doing the thing for real, even to the extent of making up a pre-production prototype. For this reason, I have not given feasibility studies a section of their own, but set them here close to the end of the book.

There are commercial as well as technical considerations to a feasibility study. If nobody wants to buy one of your works of genius, then it will sit on the shelf gathering dust unless you make use of it yourself.

Technically, experience may tell you that some things are possible. For instance, a requirement for a second order filter with a cut-off frequency of 1kHz or a ×10 amplifier are both possible in principle. We just have to take a standard circuit configuration, pummel our brains a bit over the circuit values, then arrange for its incorporation in an enclosure or as part of something greater.

Other experiences may tell us that something is possible but beyond our competence; very few of us would feel comfortable, I suspect, if faced with the task of putting together a cellular radio telephone from scratch. Yet other experiences may tell us that something is possible but beyond our resources; supercomputers are possible but expensive, and however good we are at programming, how many years will it take to write the operating system?

Where experience does not provide big clues like this, a feasibility study must be used. Essentially a feasibility study is just like embarking on the project itself, with an agreed point at which a review will be made, or with agreed criteria for termination of the study.

For instance, the project may be reviewed after the block diagram stage is reached, with the intention of seeing whether or not there are any blocks whose implementation is going to give problems. Or it may be that a rough parts list is to be looked at to find out whether costs are going to be prohibitive; or maybe there is a limiting factor on size or weight which must be addressed ahead of all other considerations.

Other things which a feasibility study might address are the purely commercial considerations of finance and expected returns. The technical nature of the product impacts on this aspect, since differing production methods have differing costs and benefits. These will be discussed at greater length later.

Another commercial issue which has its roots in the technical specification of the product is that of performance, as related to competing technologies or products. If we cannot produce a faster, lighter or stronger widget then we cannot compete in the market and perhaps should not go ahead, unless the market is expanding and we can take a share from the competition. Performance issues *per se* are rooted firmly in technical bed-rock and can only be decided in the laboratory, not the boardroom.

Feasibility studies prevent the project from progressing to the point where a 'project-killer' situation develops in which the project founders in a disorganised way (usually with some unwarranted expense or waste of time). A feasibility study attempts to foresee situations like that. Whether you use a feasibility study as a guide to avoiding project-killer situations or whether you use it as a reason to abandon given projects altogether is up to yourself.

In the event that you decide not to go ahead, then the work done on a feasibility study is not wasted; any conclusions can be re-examined as time goes by and used as a springboard for resurrecting this project or starting another.

Preliminary testing is a technical support tool for the kinds of decisions which feasibility studies address. Such testing may not always involve the testing of novel circuits. The success of the project may hinge upon a single feature of one component. We may need to test a diode, say, to see whether it can be used as a level sensor in a liquid nitrogen bath. If it cannot be so used, then the project is at an end, unless the study is widened to search for other components or techniques which might do the job.

We might also attempt to verify on paper that some novel subsystem works before going ahead with a prototype. For instance, a sensitive weighing machine might be possible, in which capacitance measurement determines the balance point and electrostatic attraction in the same pair of plates drives the scales to the balance point. An hour or two spent with a calculator might tell us whether such a system is possible or not. Further problems might then arise which have to do with accuracy. I do not know if such a pair of scales actually exists; a feasibility study for that is left as an exercise for the reader!

Work flow

As you read this section, keep in mind that nearly all of electronics' production is *assembly of components*. For most organizations there will be little, if any, actual *manufacture*. Only the biggest companies make their own resistors, their own transistors, their own switches, or their own liquid crystal displays. Such items are technically and financially (investment-wise, that is) beyond us, perhaps, or just uneconomic to make ourselves; so we buy them instead.

What manufacture there is can be farmed out, in many cases. The greater the quantity that needs to be manufactured, the better the price these items may be had at and the greater the benefits seen by the organization. The manufacturing examples commonly performed in-house which spring immediately to mind are those of p.c.b. manufacture and panel and enclosure preparation. The business of outside manufacture is looked at more closely below.

The flow of materials is worth thinking about as soon as production of more than one item is considered. In an ideal world, a loose tray of components, generated in the stock room, is gradually transformed into a product, packed, then shipped to the customer.

At the very least, to promote good work flow you should separate areas intended for soldering and assembly from those areas where stocks of parts are kept, where metalwork is under way, and those areas where administration, design and general paperwork are done.

This does not imply that a brick wall needs to separate these domains or that they should become personal empires or enclaves; in small or quiet factories a simple lane marker will do. The important things are: the flow of work from one point to another should be as smooth as possible; work should not need to be diverted from its normal course; warehousing and administrative work should not interfere with production by getting physically in the way.

There are three possible scenarios for work flow within the factory space itself. Which of these you choose depends very much upon the size and complexity of the item being produced.

First, you might be making something so simple that it would be a waste of time and effort to transfer it along a production line. Construction, assembly and testing can be done by one person or a small team working in one given area. There may be several such production cells, perhaps, each equipped with all that the team needs to do the job; they may even be working on different products.

If you're just starting up with smaller quantities of your first product then this is probably how you'll be organized anyway: one bench, with component storage at one end, a soldering iron in the middle and some test gear at the other end.

Second, larger items or those which demand special assembly or testing might benefit from more of a production line approach. Where a lot of time is invested in each article, it is worth thinking about a production line. Where expensive equipment is involved, it makes sense to use one set to serve all purposes.

Essential but expensive gear can be most usefully used by keeping it running all the time and using it to provide a centralized service for the rest of the organization.

For example, you won't have invested in a plated through hole (p.t.h.) p.c.b. line for every team. Instead, there will be one p.c.b. facility, staffed by people whose sole responsibility (at least for the times when they are thus occupied) is to serve the rest of the organization in that manner. Generally, the results of p.c.b. production (and any other common sub-assembly work) will be routed to a stores location whence it can be picked off for distribution to the assembly sites.

Where vast quantities (say thousands per week) of an item are made then a production line approach becomes almost *de rigeur*, as otherwise the problems of work flow threaten to swamp everything else. Where you see quantities of items bunching up waiting to be processed further, this should raise questions about streamlining the process, unless you're deliberately batching-up items.

Of course there are no hard and fast rules here and you may discover that some processes (testing or the making of certain sub-assemblies, for example) are best carried out by a specialist team which serves the rest of the organization, with assembly effort diffused into a number of work cells. We may then have a system which is a production line at the outset, which becomes work cell oriented for assembly and which then reverts to a production line for testing. Any combination is possible.

At the large end of the scale, we are back to the situation where it is less likely that we will want to move things around the factory, the reasons being those of size and weight this time rather than simplicity. Instead of the article under construction moving through the factory, the factory moves past the article. Bulky control cabinets, for instance, can stay where they are while production teams work on them in work cell fashion. Meanwhile, specialist teams visit the different areas to work on certain aspects with the specialized equipment and knowledge at their disposal.

Where work moves through the factory, it should be accompanied by some documentation in order to track its progress. Although such a system could be used, I suppose, to compare the performance of different assembly teams or to check up on some individual who is thought to be slacking, its purpose is actually more benign than that; it is intended just to provide a guide or a reminder as to what work needs to be done to complete a certain phase of the job.

Where jobs are passed from person to person, team to team, or shift to shift, there must be some way of telling the receiver that all the work up to now has indeed been done and that they may continue with the work in the full knowledge that their efforts are not all in vain. Tracking of this kind can help enormously where there are variants and customizations to be carried out, too.

To assist in this tracking effort, apply a serial number right at the beginning or during preparation of the enclosure. Unless two lines of work are to marry at some point, i.e. an anonymous circuit board is to be fitted to an anonymous enclosure, then keep everything together in tray or box or on a trolley along with the work sheet until assembly reaches a stage where there is only one 'lump'. Sub-assemblies can be treated in the same way, with a serial number being applied before any customization takes place. In this way, customized boards can be identified immediately by referring to the serial number.

The sheet which accompanies the article should also be marked with the serial number so that the article and the sheet can be put back together if they ever become separated. Such a document should also bear any information about standard customizations and which components must be changed in order to achieve customization.

The idea of 'standard customizations' might seem perverse, however there are a number of situations where this might arise. For example, an amplifier might be able to be customized as to gain; a filter might need to be customized in terms of its cut-off frequency. Customers could, for a small premium, order particular gains and cutoff frequencies not in the catalogue and have their requirements easily met this way.

The sheets for true 'specials' will probably have supporting documentation attached rather than the tick-the-box approach which can support standard customizations.

The laying on of hands: storage and tidiness

There was a time, when you were working on the kitchen table, when you could afford to lose something for half an hour and the time spent looking for it was an irritation but nothing more. By the time full-scale production is in swing, those days are over. Not only is personal tidiness at a premium, with other people making demands for information at odd moments, but items of stock must be easily found and used. Lost stock costs money; you might as well not have bought it in the first place.

Try and avoid keeping things in bins marked 'miscellaneous' and try and avoid pokey corners. Components and trash will mix and accumulate in these areas and you'll suffer for it.

Train your stores people, if you have specialists in that area, to recognise electronics parts for what they are and to be alert for possible mistakes in stores requisitions. There's nothing more irritating than a storeman who recognizes nothing but the part number and persists in giving out the wrong part because no-one seems to know what the right part number is and it's been put in the wrong bin anyway.

In the work cell situation, there may be a small stock of components sited locally. These must be stored in a way which prevents them from getting shuffled or lost and proper arrangements must also be made to replenish them regularly from the central store.

Whatever happens, you must have a mechanism for determining stock levels and for replacing them in good time. An open-loop system whereby remaining stocks are calculated by measuring their use during production is satisfactory, provided no stock is wasted or used in a way which bypasses production. A closed-loop system which involves counting the stock gives a truer picture of reality, but involves a lot of work, which is why this counting or 'stock-taking' is only done once a year in many organizations.

'Just in time' stocking is worth considering as it results in less stock in hand and therefore less money laid out, but it represents a good deal of planning and is very much a commitment to produce and to continue producing.

As well as finding cash for the prototype development, you must fund 'work-in-progress'. Basically, the longer something is lying on the bench the more it costs you in terms of money laid out. Minimize work-in-progress. Things should not lie about in vast quantities, unprocessed, unless batches must be big. Stock lying in warehouses is not doing a thing for you except providing an emergency back-up if deliveries fail. Your supplier stocks components, not you. Conversely, you stock for your customer (unless your customer is an outlet themselves whose business is that of stockholder).

The material resources for development may actually take up more room than the construction effort for small projects, but this situation can be rapidly reversed once a few are being made as a batch. Allow space for work-in-progress. Estimate the amount of space needed for keeping partly completed products, given that they take a certain time to complete and that there are a given number in the system at any one time.

Access is important too, not just space but the ability to gain all round access to those projects which are just a little too large to manhandle a lot. A Lazy Susan-type turntable is not a bad plan in some circumstances.

Production methods and finance

There are basically two ways to produce. There is the do-it-yourself (DIY) system where everything is done in-house, and there is the sub-contracting method where custom parts are made for you by specialist firms. Usually, technical and financial constraints will dictate a mixture of these.

As an extreme, you could have everything in the way of manufacturing and assembly done by contractors and just act as a design office, sales room and distribution outlet; even those functions, or parts of them, can be taken over by outside contractors, too. There are many firms who have successfully gone down this road; but there is the danger of losing control.

As the organization grows, the DIY elements will be replaced in some areas by custom parts bought in. Following on from this, there may be a return to DIY but with automation featuring heavily. For the smaller and less expensive kinds of production machinery, there may be a direct jump from 'craft-DIY' to 'automated-DIY' without the intervening step of contracting out. In the following paragraphs, I've attempted to summarize the technical and financial considerations for each of these situations.

For small production runs, it may be possible to do everything in-house. On the other hand, you may have no choice, that is, you may be constrained to adopting a purely craft-DIY approach.

For individuals and very small organizations, there may not be the finance available to take advantage of specialist outside contractors, and little likelihood of borrowing to get the finance together. The reason for this reluctance on the part of lenders (as I now come to understand it) to lend to the smaller organization is that the small sums involved are unlikely to generate a cash flow sufficient to service the debt. Never approach a lender with a plan to borrow money to enable you to just carry on doing what you're doing at the moment!

There may be technical constraints which dictate the use of some special component, but, generally speaking, if you do not intend to make many of a given article, then avoid the need for custom parts. If you're in this position, then ordinary manual tools and ordinary craft skills will have to do.

There will come a point, after a certain level of production is under way, where it becomes sensible to think about sending out to specialist contractors for some aspects of production. This will usually happen at the point when the sheer weight of work begins to take its toll, alerting the poor proprietor to the fact that he needs to get some sleep!

A little planning goes a long way here. Not only ought it to be possible to forecast the point beyond which it is economic to go outside, but by incorporating certain features into the design at the outset, if you have the resources, there will be a smoother transition from DIY to contracting out. Keeping documentation in good order means that it can be supplied to a contractor, in the form of a working document, as soon as the need for his services is identified.

Even if the cash is available to splurge on the latest automated widget-wrangler in-house, don't be tempted to spend unless the acquisition is going to generate cash flow. To do anything else is a waste of money; use contractors until you can justify an acquisition of this magnitude properly. As well as saving you the capital cost of acquiring production systems, specialist firms have a wealth of experience which you can easily tap into.

Incidentally, the things which seem to be most popular for farming out to contractors are panels (and enclosures generally), printed circuit boards, and any custom transformers or inductors. You can also hire specialized test equipment and development systems.

Since the printed circuit is in such common circulation and is technically difficult to make, at least for boards with more than a single side, it is usually the first item to be placed with an outside contractor.

Coming back full circle, we come back to in-house production but with automation much in evidence (the stage which I have dubbed 'automated-DIY'). Where it would be impossible to justify the acquisition of large plant and machinery for small or modest runs, it becomes sensible for larger quantities. Borrowing money becomes a real option at this stage, not least since the lenders can identify a real object which their money has bought!

So how do we set about deciding on a strategy or justifying the financial outlay on machinery?

First, the technical considerations. Wire wrapping and stripboard do not make sense beyond a few pieces, since they cannot be sensibly automated. Also, the transition from a few pieces per year to a few tens of pieces, which implies a move from craft-DIY to contracting out, means a re-design, if our original plans centred on either of these two construction methods.

These methods are time-consuming not only from the point of view of the actual work itself but also from the point of view of the need to rectify mistakes made during assembly. The speed with which a p.c.b. can be populated, as compared with either of the other methods, leaves them standing in the dust.

How does this work out in terms of actual figures? A plausible time to complete the wire wrapping for a board might be four hours. The cost of the assembly work is then, say, forty pounds. Manually assembling a p.c.b. of similar complexity would probably take a quarter of an hour. Even if we are generous and say half an hour for the p.c.b., then this assembly work costs only five pounds.

There is the investment in the p.c.b. design to be considered; let us say that the p.c.b. cost us a thousand pounds to develop, having taken our top engineer a fortnight's work. Furthermore, each p.c.b. costs us thirty pounds to buy from our supplier, for ten off, but only fifteen pounds apiece for a hundred off.

So making ten of the wire-wrapped boards costs us £400, whereas ten of the p.c.b.s costs £1350. A definite case for staying with wire wrap. But when we move over to the hundred off, it costs us £4,000 to assemble the wire-wrapped boards and only £3,000 (100 × 5 + 1000 + 1500) to assemble the p.c.b.s. That's £1,000 saved, provided that we had, or could borrow, the original £2,500 to invest in p.c.b. development costs and purchase of the boards.

We have the happy option, then, of hanging onto these profits against leaner times, passing the savings onto the customer to undercut the competition and increase sales, of ploughing the money back into further projects, or of giving our loyal workers a nice bonus (which could amount to ten pounds per board made). In reality, I suppose that a mix of these is the norm.

This calculation is a little simplistic, since it doesn't take into account the differences in the material needed by the different methods. In particular, wire-wrap designs, sockets and boards do not come for free! We really should consider the *difference* in costs between wire wrapping and p.c.b. manufacture. Also, I haven't taken into account inflation or net present values (NPV); see below for an explanation of this.

I've loaded the poor p.c.b. up a bit in order to exaggerate the differences. Even so, it demonstrates amply the advantages of making a quick calculation before committing to one technology or another.

As a rule, manually loading and soldering a 14-pin d.i.l. i.c., or a socket for the same, should take no more than thirty seconds, whereas wrapping all fourteen pins will take a few minutes, since there are all the cutting, stripping and placement actions to be done. It's up to you to perform the calculations necessary for your own pet project to see whether it justifies a p.c.b.

For large numbers of articles, specialized production tools will speed up the process of assembly. As we move away from the tens of items per week towards the hundreds, special production tools start to look even more attractive. Retaining slow manual methods is short-sighted if your time or the time of your staff is worth anything at all.

To justify the acquisition of a piece of machinery, it must first of all generate a cash flow. That is to say, part of the revenue which the organization generates must be attributable to having bought the machine. If we cannot attribute a cash flow to the purchase, then what was the point of the expense?

This cash flow might be in the form of a saving due to the removal of some other outmoded process or it might be due to the ability to do something which was not previously possible. It's not always possible to predict exactly how well the new machine is going to help us, especially in the latter case, but we can estimate the likely sales and cross our fingers (it would be a wonderful world if we could guarantee sales, especially as a start-up).

We could do the same thing by putting in a 'load factor', an estimate of how much use the machine will see as compared with how much use it could potentially see. For instance, the new p.t.h. line could produce a million boards per year, but we're going to run it at a quarter of a million, so the load factor is 25% (this may seem rather low; in fact, load factors like this can be justified in the short term if a ramp-up in production is anti-cipated and the costs of conversion are high).

The net present value (NPV) is the value of our investment at today's prices, taking into account inflation. The cash flow from the machine under consideration needs to be discounted over the expected life of the machine in order to get a true picture of its worth. In fact, the cash flow from anything needs to be discounted in this way to get its true value. In the first year, the cash flow is probably negative; this will be the case if the machine costs us more than the revenue it generates during one year.

There will be a break-even point, at which time the machine has just paid for itself, and there will be an accumulated discounted revenue over the life of the machine. These figures will necessarily change as the load factor, expected life, discount rate and so on are altered. Using a spreadsheet to play with this figures is a useful exercise. Table 13.1 shows such a sheet.

Year	Items	Price	Costs	Cash flow	PV	Cum. PV
1	500	90	50,000	-5,000	-5,000	-5,000
2	505	90	15,000	30,450	27,710	22,710
3	510	90	16,500	29,405	24,350	47,059
4	515	90	18,150	28,214	21,261	68,320
5	520	100	19,965	32,065	21,989	90,309
6	526	100	21,962	30,589	19,089	109,397
7	531	100	24,158	28,918	16,422	125,819
8	536	100	26,573	27,033	13,970	139,789
9	541	105	29,231	27,619	12,988	152,777
10	547	105	32,154	25,265	10,811	163,589
11	552	105	35,369	22,623	8,810	172,399
12	558	105	38,906	19,666	6,969	**179,368**
Discount rate				9%		

Table 13.1: Example NPV spreadsheet

In this case, the machine will cost £50,000 and will cost £15,000 to maintain in the first year, this figure rising steadily as the years progress. I've assumed that we start producing at 500 units p.a. and that this output rises modestly at about 2% p.a. The price rises in jumps every four years.

What can we conclude from this? We will lose money during the first year, but will have made a profit during the second. We're still in profit in year 12, but the profits are rather small, the machine having cost us £39,000 to make £7,000 at present-day values. If the profits get any lower, it will start to be more worthwhile leaving the money in the building society.

Overall, however, we have a nice profit of about £180,000, which might be enough to pay someone's salary over that time. This is better than the building society, which might yield about £70,000 over the same period *at present-day values.*

A spreadsheet like this need not only model a single machine. Similar tactics can be used to predict the outcome of setting up an entire factory or a complete business.

Remember that there is a ready market in second-hand automated assembly equipment of various kinds. Unless maintenance costs are going to skyrocket on an old machine, it's worth thinking about the second-user market.

On a more general note, if you can arrange cash on delivery for your sales then fine. For small runs or one-offs ask your customer to put part of the money up front. A track record or reputation helps here of course! An advance generates a tax point, by the way, and your customer can ask you to generate an invoice for the advance; probably the best policy anyway.

The foregoing says nothing about time to market. In fact, time to market has been reliably demonstrated to affect the ultimate profit over the lifetime of the product. By missing the boat, as it were, profits are lost, although like most things it is difficult to quantify by exactly how much. Again, a spreadsheet will help you to see what's likely to happen.

One final word of advice; get out of the jobbing shop mode as soon as you can. It's hard work, although some of us enjoy it for its own sake. Full-scale production, once set up and running smoothly, is the thing which creates jobs and generates revenue.

It's worth taking a closer look now at one or two of the specialized technologies which are available from outside contractors.

As for marking panels, a self-adhesive label (of sufficiently high quality) printed with vinyl ink or similar can look great. If you're having panels made outside, you may find that the firm that does them will mark them up with the required lettering and diagrams too. To go one step further, there are firms which produce flat panels which embody any necessary membrane keypad switches and which can even have windows for displays. These just stick onto your existing metal front panel.

If you're making a lot of items with a plastic case and you're sure there isn't an off-the-shelf type then consider having a custom plastic moulding made. Although tooling costs are expensive, the per-piece cost is low once the presses are rolling. A properly designed custom enclosure can look stylish or classic (or both) and helps to distinguish your product from the also-rans.

I was once investigating the cost of such a moulding and was told that there was no charge for the materials, and that one 12-hour shift (including cleaning up) would get me six months' worth of production! There's no way, financially, that the purchase of such a machine in-house could be justified, just to run for twelve hours every six months!

Polyurethane foam cored mouldings are promoted as being inexpensive too, even in smaller runs. The motto is, keep looking around for ideas and ask the people who know. Technology changes all the time. Metalworking shops, plastics moulders and printers are all keen to accommodate you and to tell you how it's done. And if they don't seem to be keen, then there are plenty more where they came from.

Testing: is it safe, does it work?

It's easy to rework a one-off, especially if you've laid in the basic ground rules for maintainability and constructibility, but reworking a few hundred units, even if you have not suffered the embarrassment of having the customer return them, is not funny.

Presumably your design is sound. But you still want to sample production to see that standards are being adhered to, even if you don't test every article. I don't like the idea that the customer tests the equipment. You test it in the factory before it goes out, unless you're very sure of yourself.

Testing is a whole subject in itself. It is mathematical and statistical and ripe for plucking by academic hands. I'm going to confine myself to some well worn rules of thumb.

The first thing to think of is the cost of reworking. There is the old saw about the cost pyramid: it costs ten times as much to dig a component out of a board as it does to reject the component before it is inserted; it costs tens times more than that to dig the board out of the system and then trace the faulty component; if the unit has to be returned from the customer, it costs ten times more again. These figures are very loose but they are not much of an exaggeration, when you take into account the possibility of lost sales due to a flagging reputation and the administrative and shipping costs incurred by returning units to base.

It can pay dividends, therefore, to test components on arrival, and it certainly pays to test subsystems and completed units to verify their correct operation. You can pay a small premium to have components tested before shipping; make sure your contract with such a supplier has a penalty clause just in case a faulty component should ever turn up. Whether it is economic to pay for such a service depends upon the complexity of your product and how easy it is to rectify any faults.

There is some justification in thinking that, for small runs, the costs of reworking are low. This is because the time to rework is hidden in the development costs. For the extreme example of a one-off, it is hard to differentiate between development and reworking, thus, reworking is not seen as a problem, rather it is seen as a kind of deferred design; the engineer says 'we will check everything once we are finished'. This disguise will not wear, however, when the customer returns the unit as faulty; the distinction has been made and reworking is the only label that can be applied.

Burn-in testing, if you have the space and time, can be a boon. It sorts out the'infant mortality' problems. Components are more likely to fail when new than later in life. Manufacturing defects which do not show under a quick test might cause premature failure under the stress of actual operating conditions. Those components which survive infancy are likely to have a long service life, until old age starts to pick them off again. The pre-production prototype should be burnt in anyway.

I remember one Japanese television manufacturer which was most reluctant to set up shop on UK shores since it would not tolerate the 1% failure rate quoted by the local resistor manufacturer, who thought that 1% was fine. It doesn't sound like a lot, after all. If you think about it, a set must contain a few tens of resistors or perhaps even a hundred, unless it is one of the really modern, highly-integrated ones.

I can't remember who gave in in the end, but my sympathies lay with the television manufacturer. It's easy to work out how many sets will still be working if you incorporate a given number of resistors.

If you have only one such resistor in your product, one percent of units will fail due to that resistor. Looking at it another way, 99% will work, and any individual unit will have a 0.99 probability of working. If you have two resistors in the set then the probability of them both working (and the set working) is $0.99 \times 0.99 = 0.99^2 = 0.9801$.

Similarly for ten resistors the probability of them all working is only $0.99^{10} = 0.9044$ (nearly 10% of assembled units are likely to be non-working at manufacture). At twenty resistors it's 0.8179 and at fifty it's 0.6050. Nearly 40% of your product is likely to be shot if it contains 50 of these resistors!

The other side of this Japanese coin, incidentally, is that the manufacturer takes responsibility for the product until the last customer gives up on it and it finally goes on the scrap heap. Try to avoid adopting the attitude that once it's sold it's forgotten, because that's a poor way to engender customer loyalty.

Yield is not the same as the probability of early failure. Take, for instance, colour liquid crystal displays for laptop computers. The yields are rising now, but they were rumoured to be in the low tens of percent, if not less than 10%, at one time (actual yield figures were secret then). Abominably low yields. Perhaps 90% of production was scrapped. However, to the outside world, all appeared well, apart from the fact that the prices were high; the manufacturer just didn't let the duff ones go out of the door. Simple, *neh*?

While we're on the subject of reliability, it's worth exploding some myths about mean time to failure (MTTF). In any reasonable sample you always get a few units which are so long lived that they extend the average failure time. The fact is, that 63% of units will have failed by the time the MTTF is reached. So MTTF really means 'the time by which most units will have failed' whereas it is often taken to imply 'the time at which most units start to fail'.

MTBF, incidentally, means mean time between failures, and implies that repair is possible; the mean time to repair (MTTR) must be added to the MTTF to get the MTBF.

To summarize, then: if it doesn't work to the specification, it doesn't go out of the door. Test and uncover mistakes. Contain your mistakes and rectify them. Rectify not only by fire-fighting the effects but by making the changes which will make those particular mistakes unlikely to happen again. We all make them, but by containing them and rectifying them we might not appear to.

Above all, don't try to make mistakes in planning look like mistakes on the shop floor. It doesn't wear well.

Staff training, policy and legislation

Training is important. How often do we hear this lip service? There seems to be an attitude that somehow we'll jog along and muddle through without training and that we'll pick it all up as we go along.

Anyone unfortunate enough to be in this position will just pick up enough to get by on a day-to-day basis. If any situation should arise which is not strictly within the bounds of previous experience, then the person is left guessing. The net result is avoidable mistakes and wasted effort trying to guess the result of taking one action or another.

I would hesitate to suggest that the foregoing is the result of a deliberate policy. In fact it reflects a poor or non-existent policy on the part of management. I don't just mean training policy; a boss who can't explain the operation of his own business sufficiently well to get it across to hirelings is a bit of a menace.

It seems to me that this lack of good policy is bound up with the 'entrepreneurial crisis' which befalls the one-man band when on the point of first hiring staff. If you're in this position then you need to be extra careful to provide a firm welcome to the people you hire. Don't treat them as a simple extension of your own personality or leave them to fester in ignorance. Decide how much a person needs to know to start the job effectively and give them that information, and the time to assimilate it, when they first arrive. Define their role fairly rigorously too. If they then grow into other roles as time goes on, then fine.

There is a place for formal training but there is no need to be formal on all occasions. Five minutes invested in explanation may save a lot of wasted time in going over the same work twice, but it needs to be said *before* the situation arises for it to count as 'training'. If you need to explain the same thing several times to several people then you have a candidate for a more formal approach.

Lack of good policy may also stem from the fear that people 'lower down on the hierarchy' may somehow take over one's job. Jockeying for position by the insecure may be tolerable in huge organizations with the muscle to carry the dead weight, but it's not something to be encouraged. Instead, encourage a secure feeling from the start by being open with policies and demonstrating that you have a grip of what's going on.

Develop a written policy, starting out by covering the run-of-the-mill things. Using this guide, your staff can make the same kinds of decisions that you would have made had you been there in person. This relieves you of doing everything yourself, and that's what you hire people for, isn't it?

Developing policy demands commitment, of course. It makes you think about what you're doing and it may also make it more difficult for you to change your mind on the spur of the moment or make arbitrary decisions. People who think on their feet abhor policy, but if an organization is to function well, rather than lurching from one chaotic situation to another, then it must have policy.

Avoid being vague, but beware also of writing too much into a policy. Leave alone those things which are none of your business or which do not matter one way or the other. Eventually, you can ask your trusted staff to develop policy for you. You will still need to cast your eye over it to see that it fits the law, the organization's image and the organization's goals.

Policy is not cast in steel. Policy is up for change but that it must be discussed and agreed, or at the very least communicated to those who need to know. I remember being carpeted once for using a certain programming language to develop some software. It turned out that use of BASIC was the firm's policy, although neither myself nor anyone else in the organization (including the person who thought it up) knew of it until that moment.

Everyone has gaps in their knowledge and it always takes time for someone to find their way when new on the job. Don't expect that you're just buying in a skills package when you hire staff; help people to settle in by giving them an introduction to the organization and their part in it, formal or otherwise.

Train all those who are willing and capable to cover as many aspects of their work as they can, not from the point of view of making do with less people, but more from the point of view of fostering participation and involvement.

In our culture (I'm sure this is not the case for some others) no-one likes to think of themselves as just a cog in a big machine. Multi-skilling, coupled with an open and accessible policy, provides a person with an insight into the workings of the whole organization and makes them capable of far better integration with their colleagues.

Most technical legislation, so far as electricity is concerned, has to do with preventing people from coming into contact with voltages sufficiently high to cause a shock. Apart from this, the most important forms of legislation are to do with safety at work and product liability.

My own feelings on the matter are that if you have given sufficient attention to detail and if you are committed to quality, very much of what is contained in the legislation will have been amply covered. There are one or two rather bureaucratic items which need to be complied with, such as the need to display the various acts of parliament on the premises, but these are not onerous.

Going one better than the law, rather than attempting to skate round it, is a good policy. Seriously, you should regard the requirements of the law as a lowest common denominator, rather than as a high standard to be achieved, especially where safety is concerned.

At one engineering firm in which I worked, the Health & Safety guys came round once. They seemed happy about everything else but gave us a warning to wash our hands if we'd been doing any soldering, especially before eating anything. Apparently lead poisoning is a problem. The less exotic electronics processes use very little that is toxic, but I suppose solder is an exception. It's a point worth remembering.

Legislation apart, employers who genuinely care about their employees and who are seen to care will have few problems with staff or labour relations.

14: Further Reading

Here is a list of some of the books which I have found particularly interesting or informative. Some of them I've known for a long time, others are recent finds. Most of them are represented on my own bookshelf, but not all. Mostly they examine some particular aspect of electronics in depth and most of them provide practical hints and tips on getting real working circuits going.

Some of them are supplier's catalogues, which can be just as informative. I have *not* included catalogues which are *just* lists and no more.

One or two of them are rather long in the tooth now, but I have selected them for their timelessness; they are all just as valuable now as the time when they came into print. Others are books which I've only come across recently but which have left their impression.

Active-Filter Cookbook, Don Lancaster, Sams

I suppose this was my first true love in electronics books. I've lent it to numerous people but, unlike many things which are lent, it always finds its way home. I've no doubt that it will be out of print by now, unfortunately. It is a truly comprehensive guide to filters using op-amps. It has the rare capacity to touch on both the practical and the mathematical; you can read it without the math if you want, but it's there to back you up if you need it.

Art of Electronics, Horowitz and Hill, Cambridge

This, I suspect, is the one we all wish we'd written. Thoroughly comprehensive, covering most aspects of analogue and digital design, and crammed with examples. This is a designer's guide, and so stops short at full-scale production and related issues.

Circuit Designer's Companion, Tim Williams, Butterworth Heinemann

This is a good general romp through circuit design which includes one or two items not found in run-of-the-mill textbooks. There are some good general tips on electromagnetic compatibility and other, more mechanical, aspects of electronics.

CMOS Data Book, National Semiconductor

A comprehensive guide. If I'm doing anything with CMOS, mine is never shut. In fact, the poor thing hardly ever sees the bookshelf!

Coil Design and Construction Manual, B.B. Babani, Tandy

This is the only book in which I've come across really detailed instructions on the design and making of inductors. Covers everything from r.f. to d.c. There are some useful wire tables in the back too. If you must design or make your own inductors, then this is handy.

Industrial Electronics, Noel Morris, McGraw-Hill

This book is particularly strong on discrete transistor circuits. If you're into designing from scratch and don't mind some fairly heavy algebra, then you will find this very useful.

Linear Data Book, National Semiconductor

Just the CMOS data book mentioned above, this is comprehensive. Again, I wouldn't be without my copy, dated as it is. I also have the Linear Applications Book, which is useful enough, but I still like to root around in the actual numbers too.

Maplin catalogue, Maplin Electronics plc

Occasional in-depth explanations unusual in a catalogue, but extremely useful. One or two components which are hard to get elsewhere. Some good kits!

Microcomputer interfacing, Howard Stone, Addison Wesley

Covers computer interfacing in all its respects, memory, peripherals and buses. In particular, there are useful explanations of transmission line theory as applicable to the computer bus and of earthing and ground loop problems.

Newnes Electronics Assembly Handbook, Keith Brindley, Butterworth Heinemann

This is a detailed account of manufacturing technology related to electronics. If you're contemplating automation, then this book is highly recommended. It also lists relevant standards publications and has a complete chapter on reliability and testing.

Practical Electronics Calculations and Formulæ *and* Further Practical Electronics Calculations and Formulæ, FA Wilson, Bernard Babani

A pair of little gems crammed with all kinds of numerical goodies. Not for reading, but for keeping by you and dipping into when needed. In total, 700 pages of formulae to calculate your way across every aspect of electronic circuitry. Some explanatory passages with generally good introductions to topics as well.

RS catalogue, RS Components

So large now that it spans three volumes. A comprehensive overview of everything that's commonly available, with pictures so that you can see what everything looks like, and brief explanations where needed. RS also publish their own data and applications sheets. Individuals as well as companies can obtain a catalogue from Electromail, the cash-with-order subsidiary of the Electrocomponents Group, who use the same catalogue.

Tower's International Transistor Selector, TD Towers, Foulsham

Well known and probably the best source of discrete transistor information. The current edition is version 4. Very comprehensive and accurate, although I did once catch them out with a pinout or two. There are others in the series too, covering op-amps and so forth. In the front of the op-amp version is one of the most lucid and practical explanations of op-amp behaviour I have seen so far.

TTL Data Book, Texas Instruments

Last but by no means least. Everything you always wanted to know about TTL. If the recent innovations in logic symbols and nomenclature confuse you, more recent editions of this book carry a full explanation.

INDEX

D

E

F

M

N

O

P

Q

R

S